GENERAL INVESTIGATIONS OF CURVED SURFACES

CARL FRIEDRICH GAUSS

Edited with a new Introduction and Notes by

PETER PESIC

St. John's College, Santa Fe, New Mexico

DOVER PUBLICATIONS, INC.

Mineola, New York

Copyright

Bibliographical Note

This Dover edition, first published in 2005, is an unabridged republication of the work first published by the Princeton University Library, Princeton, New Jersey in 1902. It was translated from the Latin and German by Adam Hiltebeitel and James Morehead. A new Introduction, Additional Notes on the text, Appendix, and Bibliography have been prepared by Peter Pesic.

The editor wishes to thank Roger Cooke, Grant Franks, and Jeremy Gray for their helpful comments.

International Standard Book Number: 0-486-44645-X

Manufactured in the United States of America
Dover Publications, Inc., 31 East 2nd Street, Mineola, N.Y. 11501

INTRODUCTION TO THE DOVER EDITION

"Thou, Nature, art my goddess; to thy law my services are bound." By choosing this motto, Carl Friedrich Gauss (1777-1855) indicated his deep commitment to natural science.[1] His profound discoveries earned him the title of "prince of mathematicians"; his multifaceted achievements challenge the common distinction between "pure" and "applied" mathematics. James Clerk Maxwell thought Gauss's work on magnetism "may be taken as models of physical research by all those who are engaged in the measurement of any of the forces in nature."[2] Gauss was also deeply involved in astronomy, surveying, and geodetic mapping. These intensely practical activities converged with his "pure" concerns in his seminal formulation of the theory of curved surfaces. Gauss's work, generalized by his student Georg Friedrich Bernhard Riemann, was part of the essential groundwork for Albert Einstein's general theory of relativity. According to Einstein, "the importance of Gauss for the development of modern physical theory and especially for the mathematical fundamentals of the theory of relativity is overwhelming indeed; ... If he had not created his geometry of surfaces, which served Riemann as a basis, it is scarcely conceivable that anyone else would have discovered it."[3]

From the beginning, Gauss's mind spanned practice and speculation. The son of a day laborer and an almost illiterate mother, he was a calculating prodigy who at age ten grasped how to sum the first hundred integers without tedious addition. At age fifteen, he calculated the number of primes less than a given number (going up into the hundred thousands), from which he conjectured the celebrated prime number theorem.[4] In early youth he also began considering the consequences for geometry were the parallel postulate of Euclid not true.[5] At age eighteen he used algebra in order to accomplish a geometrical *tour de force*, the construction of a regular seventeen-sided polygon using only straightedge and compass, which decided him to pursue mathematics rather than philology.

Gauss's involvement with geometry was also practical. He remained an active astronomical observer over the course of fifty years, using sextant and telescope and making lengthy calculations. His determination of the orbit of the planetoid Ceres led to his appointment as professor of astronomy and director of the Göttingen observatory. He brought together his knowledge of mathematical and practical astronomy in his *Theory of the Motion of the Heavenly Bodies* (1809).[6]

Gauss's involvement in surveying and geodesy was no less intense.[7] In post-Napoleonic Europe, economic and military considerations called for greatly improved maps, which required much more accurate surveys. For more than a decade (1818-1832), most of Gauss's time was devoted to directing the surveying of the Electorate of Hanover, a major project that involved extensive fieldwork to make triangulations, followed by endless calculations. Gauss took charge of the arduous work personally. In summer, he spent most of his time away from home, going from village to village, arranging for sightlines by removing trees, and directing his military assistants. Once he was thrown from his horse, another time his carriage overturned and his theodolite fell on him; he suffered from injuries, heat, and lack of sleep. During the same time, his wife was seriously ill (she died in 1831) and he became estranged from one of his sons, who left for America. Gauss struggled with every aspect of the observations as well as the exacting calculations they entailed. He was in fact the founder of modern geodesy. Gauss was especially proud of having invented the heliotrope, an instrument whose movable mirror could reflect light over longer distances than ever before practicable, which he devised after being bothered by the reflection of sunlight from a distant window.

Though one of his biographers deplored that so much of Gauss's valuable time was taken up by the survey, its practical needs led Gauss to several important mathematical discoveries.[8] At age seventeen, he had already devised the method of least squares to deal with astronomical data; later, he applied this fundamental tool for analyzing data to geodetic measurements.[9] The exigencies of map-making required projecting the curved surface of the earth onto a flat plane. Several projections were already in use, such as the Mercator or stereographic projections, but their varying advantages and disadvantages led Gauss to reconsider the question mathematically. In an important paper of 1822, he asked: What system of mapping would give the most faithful possible projection, in the sense that "the image is similar to the original in the smallest parts," as he put it?[10] Here emerges a deep theme: the search for *invariants*, in this case quantities that test the similarity of image to original. Because surveying involves the measurement of angles between sightlines to different points of reference, Gauss naturally considered those projections that would *preserve angles*, so that an angle surveyed on earth would be the same as the corresponding angle shown on the plane map (a condition obeyed by Mercator projections, for instance). In his 1822 paper, Gauss set forth the properties of these *conformal mappings.*[11]

The mapping problem raises the question of the relation between different sorts of maps and the curved surface they represent because different projections can render startlingly different versions of the same earth. For instance, a Mercator projection makes the polar regions seem huge in comparison to comparable areas at lower latitudes. Gauss then inquired whether curved surfaces have any properties that reflect intrinsic properties of their curvature. The surprising results form the climax of his *General Investigations of*

Curved Surfaces (Disquisitiones generales circa superficies curvas) of 1827.

In this work, Gauss begins with the perspective of a surveyor concerned with the directions of various straight lines in space he specifies through the use of an "auxiliary sphere" of unit radius, similar to the celestial sphere in astronomy. A parallel radius of the auxiliary sphere can represent a given directed line, so that the direction of that line corresponds to a certain point on the surface of the sphere. Gauss thus opens an extended comparison between spherical geometry and the actual curved surface being mapped. He then uses a two-dimensional coordinate grid on the curved surface, a construction introduced by Leonhard Euler, whose work was fundamental for all that came after.[12] These "Gaussian coordinates" (as they are now called) are far more convenient for depicting distance relations than the Cartesian x, y, and z coordinates. Here again we recall a surveyor tracking terrestrial curvature using the two dimensions of latitude and longitude. For instance, consider a point on a convex surface touched by an *osculating sphere*, a sphere that "kisses" the inside of the surface, matching the local curvature near that point. The radius of the osculating sphere is called the *radius of curvature* of the surface at that point, so that a flat plane corresponds to an infinite radius of curvature. Yet the earth is not a perfect sphere (as Newton pointed out) because rotation makes its equatorial region bulge slightly, so that the radii of curvature in the polar and equatorial directions differ over the earth's surface. In 1760, Euler showed that in this case the curvature at some point can be expressed as the reciprocal of the products of the maximum and minimum radii of curvature (signed positively or negatively to denote convexity or concavity), a definition of curvature that came to be standard.

In his quest for a truly exact geodesy, Gauss had to determine the degree to which the earth's double curvature was observable and could affect his surveys. In addressing this, he came upon far more general and surprising results. He defines the *measure of curvature* (now called the "Gaussian curvature") as the ratio between an infinitesimal area on the curved surface and the corresponding area mapped on a sphere, which is closely related to Euler's standard definition. Now Gauss took a crucial step forward. He considers all curved surfaces that can be "developed" on each other, meaning that to each point of the original surface there corresponds a single point on the map, as if the original surface were allowed to be arbitrarily bent *but without stretching*. The modern term *isometry* emphasizes that Gauss's "development" requires *equal distance* be preserved. From this, Gauss uses his surface coordinates and in a couple of lines deduces what he calls his "remarkable theorem" (*theorema egregium*): "*If a curved surface is developed upon any other surface whatever, the measure of curvature in each point remains unchanged.*"

Gauss's theorem shows that a surface can have an *intrinsic* curvature that does not change even if that surface is *extrinsically* bent (always leaving distances invariant). Thus, no plane map can be faithful (isometric) to the intrinsically curved earth, even in a small region, which may have long seemed intuitive to map-makers. Yet the beauty of Gauss's theorem is also its counterintuitive surprise; one might have thought that there is no such thing as *intrinsic* curvature, as opposed to how a surface is deployed in its ambient space. For instance, the intrinsic curvature of a flat plane and of a cylindrical surface are both zero, so that *both* are intrinsically flat, though intuitively it might seem the cylinder is curved. Here Gauss shows the irrelevance of the three-dimensional space within which we intuitively immerse the cylinder or plane.

Gauss's theorem led to a new formulation of differential geometry, the study of surfaces in the small. He introduced the use of a *metric*, an expression for the infinitesimal distance between points that generalizes the Pythagorean formula to curved surfaces expressed in arbitrary coordinates. This was the starting point Riemann took in 1854 to generalize Gauss's work to a "space" of n dimensions and determine its intrinsic curvature; indeed, Riemann thought of reformulating all geometry in intrinsic terms.[13] Gauss also developed Euler's work on *geodesics*, the shortest possible lines on a curved surface, which are the natural generalizations of straight lines in the Euclidean plane. Thus, the Cartesian grid of orthogonal straight lines on the plane can be generalized to grids of geodesics on a curved surface.

Gauss then returns to his surveying and calculates the curvature of the surface in terms of the metric (the "Gauss equation," as it is now called). He also constructs a triangle whose sides are geodesics. He shows that the excess (or deficit) of the sum of its angles compared to 180° is measured by the triangle's corresponding area mapped on a sphere (now known as the Gauss-Bonnet theorem). Gauss then uses this "most elegant" theorem to produce practical formulas he can apply to surveying. Here, his calculations become more detailed than any previous work in differential geometry, a direct effect (I believe) of Gauss's preoccupation with the masses of geodetic data he collected, over one million data points in all. Finally, he takes the largest single triangle included in his survey, between the mountains Brocken, Hohehagen, and Inselberg.[14] Gauss shows that the observed excess of the sum of these angles over 180° is only 14.85348 seconds of arc, which he calls an "insensible" difference. Thus, his practical question seems resolved: to an accuracy of about 0.0002% in angle, terrestrial surveying obeys Euclidean geometry, even taking into account the nonspherical curvature of the earth.

This would seem much ado about (almost) nothing, a laborious verification of assumptions long held by surveyors anyway. Gauss emphasizes the "new point of view" of intrinsic geometry, in which "a wide and still wholly uncultivated field is open for investigation." But he may have kept his own deepest speculations to himself. His letters show that he had not ceased thinking about non-Euclidean geometries since his youth. He told several correspondents that he had been in possession of non-Euclidean geometry decades before the earliest published works of János Bolyai (1832) and Nikolai Lobachevskii (1826), but the extent to which Gauss really worked out the full details remains dubious.[15] Still, Felix Klein thought that "Gauss still deserves the greatest credit for non-Euclidean geometry because, by the weight of his authority, he first brought this intellectual creation, heavily contested at first, to general attention and ultimately to victory."[16] In 1816, Gauss wrote that "it would even be desirable that Euclid's geometry should be false, since we would then be *a priori* in possession of a universally valid unit of length," which Johann Lambert had realized in 1766, but (unlike Gauss) took as evidence for Euclidean geometry.[17] This accords with Gauss's work in electrodynamics, for which he found a universally valid set of units still in use today.[18] Writing in 1824 about this universal unit of length, Gauss noted that "if the non-Euclidean geometry should be true, and if the constant is not altogether too large with respect to those quantities that can be obtained by our measurements of the earth or the heavens, then this constant could be determined *a posteriori*," by empirical observations on the cosmic scale.[19]

Thus, when Gauss wrote in 1825 that his work is "involved in the metaphysics of the geometry of space," he signaled that his concerns went beyond surveying. We know that he read Kant carefully and was critical of Kant's mathematical insight. We might infer that Gauss, *contra* Kant, believed that the geometry of space is probably non-Euclidean, its curvature determinable by astronomical observations. Further, Gauss judged that "geometry should be ranked not with arithmetic, which is *a priori*, but with mechanics."[20] Even though his own painstaking measurements of large terrestrial triangles agreed with Euclidean geometry, the question remains what he thought about testing non-Euclidean geometry as such.[21] The statements already quoted indicate that he was prepared to do so, at least on the scale of the heavens. At the least, his earthly surveying showed that any evidence of non-Euclidean geometry would have to be found astronomically.

Only in such private letters did Gauss disclose his radical thoughts, partly from his deep reluctance to disclose unfinished, unripe work, partly from his desire to avoid what he called "the outcry of the Boeotians [*das Geschrei der Böotier*]."[22] Who were these "yokels" (as the German student slang word *Böotier* might be rendered) whom Gauss so disdained? Perhaps there is a clue in his pointed avoidance of any political controversy, even of expressing any political opinion, throughout a long life in times of revolution and reaction. In 1837, the new king of Hanover annulled the constitution and then dismissed seven Göttingen professors who objected to his actions, including Gauss's son-in-law Heinrich Ewald and his close friend Wilhelm Weber. Unlike others who protested or even resigned, Gauss hoped that "the university as a body will not enter the political scene" and wrote to a friend "I have never told anybody what I do or do not intend to do."[23] As with his opinions about non-Euclidean geometry, here again his awareness of violent controversy and political danger moved him to keep his views private. He seemed to feel that not only revolutionary mobs but crowned heads would howl for his head were his real thoughts known. Though one might wish Gauss had had more courage to defy them, the turbulent reception of Einstein's work a full century later could be read as confirmation of Gauss's fears. The Boeotians linked Einstein's radical ideas to the "Red menace," denounced him from the pulpit, and even threatened his safety. To be sure, all this was not simply a reaction to non-Euclidean geometry; far larger movements in world history were also involved. Though the outrage that met these new ideas was great, even greater was the wonder and acclaim they inspired.

If geometry must be judged by observation, the way lies open to connecting geometry with physics and physics with geometry. Riemann and Einstein made the decisive steps towards this realm that Gauss only glimpsed. Of course, Gauss and Riemann still lacked the full equations of electrodynamics. The project of geometrizing physics could not go forward until time could be integrated with space, as Einstein learned from Maxwell's equations. The spatial position of a planet does not determine its orbit, which requires also knowing its velocity at that point, hence temporal information.[24] As Einstein acknowledged, once that has been realized, the work of Gauss and Riemann is all-important for general relativity.[25] Gaussian differential geometry remains crucial for any generalization of Einstein's theory, such as quantum gravity. In his *General Investigations of Curved Surfaces*, Gauss revealed a deeply new aspect of geometry, its inner character, revealed in its differential behavior on the finest scale.[26] As such, modern differential geometry springs from these few pages, which also delineate the inner nature of spacetime that any future physics must confront.

A word on the text presented here, which reprints the 1902 translation by James Caddall Morehead and Adam Miller Hiltebeitel, then students at Princeton University, whose original introduction by H. D. Thompson, a librarian at Princeton, detailed the history of the various versions and translations.[27] This Dover reprint follows the order of the original 1902 edition (silently correcting errata), beginning with Gauss's *General Investigations of Curved Surfaces* of 1827 (pp. 3-44 below). Then follows Gauss's abstract (pp. 45-49), which you may want to read first because it gives a helpful and non-technical overview. Morehead and Hiltebeitel's notes then follow (pp. 51-78), which tend to be bibliographical until §17, at which point they begin to expand many of the calculations of which Gauss merely gives the result. Following this, you will find Gauss's "New General Investigations of Curved Surfaces" (pp. 80-110), which, despite its title, was written in 1825, *before* the 1827 *General Investigations*, but only published in 1901, long posthumously. As such, the "New General Investigations" is an interesting earlier version of Gauss's work that allows insight into the development of his thought. Morehead and Hiltebeitel's notes to the "New General Investigations" follow (pp. 111-114), then my own notes for the whole volume, set off in square brackets [ ]. Because Morehead and Hiltebeitel's notes do not really begin until §17 of the *General Investigations*, my notes concern its earlier parts (§§1-16). My intent has been to make these sections of the work (which include some of its most striking and important results) accessible to students. To this end, I have included a brief appendix that gives some useful basic results in spherical trigonometry, along with a bibliography.[28]

When a friend wrote Gauss in 1823 regretting the loss of time he must have suffered in his geodetic work, Gauss replied that "all the measurements in the world do not outweigh one theorem by which the science of eternal truths is really carried forward. However you must not judge on the absolute, but on the relative value."[29] Gauss's *General Investigations of Curved Surfaces* show how his toil as a surveyor bore fruit in eternal truths.

Peter Pesic

[1] Gauss took his motto from *King Lear* I.ii.1-2. Was Gauss unaware that in the play these words are spoken by the bastard Edmund, claiming his "natural" birth as warrant for his crimes? It is hard to imagine he would have chosen this motto had he fully realized that this speech ends: "Now, gods, stand up for bastards."

[2] Maxwell 1954, vii.

[3] Cited in Hall 1970, 105.

[4] The prime number theorem states that $\pi(n)$, the number of primes less than n, asymptotically approaches $n/\ln n$ as n becomes very large and was proved in 1896 independently by Charles de la Vallée-Poussin and Jacques Hadamard; see Hall 1970, 10-18. The young Gauss used Johann Lambert's table of prime numbers but throughout his life Gauss "used a spare quarter of an hour to investigate a thousand numbers here and there," as he wrote in a letter of 1849 (Hall 1970, 14).

[5] In a letter to Hans Christian Schumacher (1846), Gauss stated that already in his fifteenth year (1792) he was convinced that a "perfectly consistent" geometry would ensure were the parallel postulate not assumed; see the useful selection from Gauss's correspondence on these matters in Ewald 1996 1:296-306. For the larger history, see Bonola 1955, who emphasizes the foundational question of the parallel postulate. Jeremy Gray adds the importance of analysis and differential geometry in Gray 1979 and 1989, which includes an excellent historical survey, as does Gray 2004, 3-48.

[6] Gauss 2004, discussed by Wilson 2005, 316-328.

[7] For a detailed account, see Dunnington 2004, 113-138 (still the most complete biography available), Grossmann 1955, Reich 1981, and Breitenberger 1984.

[8] Dunnington 2004, 138. In contrast, Lanczos 1965, 64-84, thanks the Hanoverian authorities for causing Gauss to undertake this geodetic work.

[9] See Bühler 1981, 137-140

[10] See Bühler 1981, 102-103. His 1822 Copenhagen Prize Essay "Allgemeine Auflösung der Aufgabe, die Theile einer gegebenen Fläche auf einer anderen gegebenen Fläche so abzubilden, dass die Abbildung dem Abgebildeten in den keinsten Theilen ähnlich wird," can be found in Gauss 1973, 4:189-216, translated in Smith 1959, 463-475.

[11] In 1772, Lambert was the first to consider in full generality the problem of conformal mapping of the sphere on the plane (Lambert 1896). For the history of the mapping problem, see Kline 1972, 2:570-571.

[12] Euler introduced these coordinates in his 1771 paper, "De solidis quorum superficiem in planum explicare licet," in Euler 1975, ser. 1, 28:161-186. For a brief review of Euler's work, see Kline 1972, 2:562-571.

[13] For a translation of Riemann's 1854 paper "On the Hypotheses that Lie at the Foundations of Geometry," see Ewald 1996 2:649-661.

[14] The sides of this triangle are 69, 85, and 106 km. Though any three points define a plane triangle, surveyors do not measure its planar angles but those lying on the curved surface of the earth.

[15] Gauss read Bolyai in 1832; after 1840, he probably read Lobachevskii's later papers in Russian but never saw his 1826 publication; see Gray's helpful commentary in Dunnington 2004, 461-467.

[16] Klein 1979, 53.
[17] See Gauss 1973, 8:186-188, and Hall 1970, 110-111. Both Adrien-Marie Legendre and Lambert thought this absolute unit was evidence of the validity of the parallel postulate.
[18] For Gaussian absolute units, see Hall 1970, 129-131.
[19] In 1807, Ferdinard Carl Schweikart and Franz Adolf Taurinus noted that deviations from Euclidean geometry might only be observable at cosmic distances, so that they termed the non-Euclidean alternative "astral geometry," a term Gauss liked and sometimes used himself. In fact, Gauss's 1824 letter is addressed to Taurinus; see Hall 1970, 111-112 and Gauss 1973, 8:180-181. For the work of Schweikart and Taurinus, see Bonola 1954, 75-83. See also Hoppe 1925, who reported the verbal recollection of Johann Benedict Listing that Gauss said that triangulation between fixed stars could test empirically the validity of Euclidean geometry.
[20] Letter to Heinrich Olbers, 1817, Gauss 1973, 8:177, cited in Weyl 1949, 133.
[21] Long controversy surrounds the notion that Gauss's triangulations were meant as tests of Euclidean geometry, argued against by Dunnington 1953; apparently unaware of this paper, the argument was reinvented by Miller 1972; see also Breitenberger 1984. Nothing Gauss explicitly wrote supports the belief that his surveying was intended to test Euclid. However, his measurements did provide information, even if negative.
[22] Letter to Friedrich Bessel, 1829, in Gauss 1973, 8:200 and Dunnington 2004, 182.
[23] For the affair of the "Göttingen seven," see Bühler 1981, 135-137; the letter cited was to Schumacher.
[24] See Adler, Bazin, and Schiffer 1975, 5-7. Rindler 1994 imagines what might have happened if Riemann had discovered general relativity.
[25] As Hermann Weyl and Élie Cartan showed in 1921-1922, Einstein's equations emerge with surprising inevitability from Riemannian geometry, given a few physically reasonable assumptions; see Pesic and Boughn 2003.
[26] For a helpful overview of modern differential geometry in relation to Gauss's work, see Dombrowski 1979, 137-153, which also includes a helpful recounting of Gauss's work in modern notation (99-120) and a discussion of its historical development (121-136). See also Coolidge 1963, 404-421.
[27] Earlier reprints include one in Dombrowski 1979, which includes only the *General Investigation* and its abstract, with the Latin text on facing pages (2-95), and the Gauss 1965 reprint of Gauss 1902, along with a brief introduction by Richard Courant.
[28] The present bibliography replaces that given in Gauss 1902, which those interested in the nineteenth-century literature may want to consult, along with the exhaustive bibliography in Merzbach 1984.
[29] This letter from Friedrich Bessel is cited in Hall 1970, 90.

HISTORICAL BACKGROUND

In 1827 Gauss presented to the Royal Society of Göttingen his important paper on the theory of surfaces, which seventy-three years afterward the eminent French geometer, who has done more than any one else to propagate these principles, characterizes as one of Gauss's chief titles to fame, and as still the most finished and useful introduction to the study of infinitesimal geometry[1]. This memoir may be called: General Investigations of Curved Surfaces, or the Paper of 1827, to distinguish it from the original draft written out in 1825, but not published until 1900. A list of the editions and translations of the Paper of 1827 follows. There are three editions in Latin, two translations into French, and two into German. The paper was originally published in Latin under the title:

I *a*. Disquisitiones generales circa superficies curvas
auctore Carolo Friderico Gauss
Societati regiæ oblatæ D. 8. Octob. 1827,

and was printed in: Commentationes societatis regiæ scientiarum Gottingensis recentiores, Commentationes classis mathematicæ. Tom. VI. (ad a. 1823–1827). Gottingæ, 1828, pages 99–146. This sixth volume is rare; so much so, indeed, that the British Museum Catalogue indicates that it is missing in that collection. With the signatures changed, and the paging changed to pages 1–50, I *a* also appears with the title page added:

I *b*. Disquisitiones generales circa superficies curvas
auctore Carolo Friderico Gauss.
Gottingæ. Typis Dieterichianis. 1828.

II. In Monge's Application de l'analyse à la géométrie, fifth edition, edited by Liouville, Paris, 1850, on pages 505–546, is a reprint, added by the Editor, in Latin under the title: Recherches sur la théorie générale des surfaces courbes; Par M. C.-F. Gauss.

---

[1] G. Darboux, Bulletin des Sciences Math. Ser. 2, vol. 24, page 278, 1900.

III*a*. A third Latin edition of this paper stands in: Gauss, Werke, Herausgegeben von der Königlichen Gesellschaft der Wissenschaften zu Göttingen, Vol. 4, Göttingen, 1873, pages 217–258, without change of the title of the original paper (I*a*).

III*b*. The same, without change, in Vol. 4 of Gauss, Werke, Zweiter Abdruck, Göttingen, 1880.

IV. A French translation was made from Liouville's edition, II, by Captain Tiburce Abadie, ancien élève de l'École Polytechnique, and appears in Nouvelles Annales de Mathématique, Vol. 11, Paris, 1852, pages 195–252, under the title: Recherches générales sur les surfaces courbes; Par M. Gauss. This latter also appears under its own title.

V*a*. Another French translation is: Recherches Générales sur les Surfaces Courbes. Par M. C.-F. Gauss, traduites en français, suivies de notes et d'études sur divers points de la Théorie des Surfaces et sur certaines classes de Courbes, par M. E. Roger, Paris, 1855.

V*b*. The same. Deuxième Édition. Grenoble (or Paris), 1870 (or 1871), 160 pages.

VI. A German translation is the first portion of the second part, namely, pages 198–232, of: Otto Böklen, Analytische Geometrie des Raumes, Zweite Auflage, Stuttgart, 1884, under the title (on page 198): Untersuchungen über die allgemeine Theorie der krummen Flächen. Von C. F. Gauss. On the title page of the book the second part stands as: Disquisitiones generales circa superficies curvas von C. F. Gauss, ins Deutsche übertragen mit Anwendungen und Zusätzen . . . .

VII*a*. A second German translation is No. 5 of Ostwald's Klassiker der exacten Wissenschaften: Allgemeine Flächentheorie (Disquisitiones generales circa superficies curvas) von Carl Friedrich Gauss, (1827). Deutsch herausgegeben von A. Wangerin. Leipzig, 1889. 62 pages.

VII*b*. The same. Zweite revidirte Auflage. Leipzig, 1900. 64 pages.

The English translation of the Paper of 1827 here given is from a copy of the original paper, I*a*; but in the preparation of the translation and the notes all the other editions, except V*a*, were at hand, and were used. The excellent edition of Professor Wangerin, VII, has been used throughout most freely for the text and notes, even when special notice of this is not made. It has been the endeavor of the translators to retain as far as possible the notation, the form and punctuation of the formulæ, and the general style of the original papers. Some changes have been made in order to conform to more recent notations, and the most important of these are mentioned in the notes.

The second paper, the translation of which is here given, is the abstract (Anzeige) which Gauss presented in German to the Royal Society of Göttingen, and which was published in the Göttingische gelehrte Anzeigen. Stück 177. Pages 1761–1768. 1827. November 5. It has been translated into English from pages 341–347 of the fourth volume of Gauss's Works. This abstract is in the nature of a note on the Paper of 1827, and is printed before the notes on that paper.

In 1901 the eighth volume of Gauss's Works appeared. This contains on pages 408–442 the paper which Gauss wrote out, but did not publish, in 1825. This paper may be called the New General Investigations of Curved Surfaces, or the Paper of 1825, to distinguish it from the Paper of 1827. The Paper of 1825 shows the manner in which many of the ideas were evolved, and while incomplete and in some cases inconsistent, nevertheless, when taken in connection with the Paper of 1827, shows the development of these ideas in the mind of Gauss. In both papers are found the method of the spherical representation, and, as types, the three important theorems: The measure of curvature is equal to the product of the reciprocals of the principal radii of curvature of the surface, The measure of curvature remains unchanged by a mere bending of the surface, The excess of the sum of the angles of a geodesic triangle is measured by the area of the corresponding triangle on the auxiliary sphere. But in the Paper of 1825 the first six sections, more than one-fifth of the whole paper, take up the consideration of theorems on curvature in a plane, as an introduction, before the ideas are used in space; whereas the Paper of 1827 takes up these ideas for space only. Moreover, while Gauss introduces the geodesic polar coordinates in the Paper of 1825, in the Paper of 1827 he uses the general coordinates, p, q, thus introducing a new method, as well as employing the principles used by Monge and others.

H. D. Thompson.

Contents

GENERAL INVESTIGATIONS

OF

CURVED SURFACES

(The notes referred to in the text are to be found on pages 51 - 78.)

1.

Investigations, in which the directions of various straight lines in space are to be considered, attain a high degree of clearness and simplicity if we employ, as an auxiliary, a sphere[1] of unit radius described about an arbitrary centre, and suppose the different points of the sphere to represent the directions of straight lines parallel to the radii ending at these points. As the position of every point in space is determined by three coordinates, that is to say, the distances of the point from three mutually perpendicular fixed planes, it is necessary to consider, first of all, the directions of the axes perpendicular to these planes. The points on the sphere, which represent these directions, we shall denote by (1), (2), (3). The distance of any one of these points from either of the other two will be a quadrant; and we shall suppose that the directions of the axes are those in which the corresponding coordinates increase.

2.

It will be advantageous to bring together here some propositions which are frequently used in questions of this kind.

I. The angle between two intersecting straight lines is measured by the arc between the points on the sphere which correspond to the directions of the lines.

II. The orientation[2] of any plane whatever can be represented by the great circle on the sphere, the plane of which is parallel to the given plane.

III. The angle between two planes is equal to the spherical angle between the great circles representing them, and, consequently, is also measured by the arc intercepted between the poles of these great circles. And, in like manner, the angle of inclination of a straight line to a plane is measured by the arc drawn from the point which corresponds to the direction of the line, perpendicular to the great circle which represents the orientation of the plane.

IV. Letting $x, y, z; x', y', z'$ denote the coordinates of two points, r the distance between them, and L the point on the sphere which represents the direction of the line drawn from the first point to the second, we shall have

$$x' = x + r\cos(1)L$$
$$y' = y + r\cos(2)L$$
$$z' = z + r\cos(3)L$$

V. From this it follows at once that, generally,[3]

$$\cos^2(1)L + \cos^2(2)L + \cos^2(3)L = 1$$

and also, if L' denote any other point on the sphere,

$$\cos(1)L \,.\, \cos(1)L' + \cos(2)L \,.\, \cos(2)L' + \cos(3)L \,.\, \cos(3)L' = \cos LL'.$$

VI. Theorem.[4] *If L, L', L'', L''' denote four points on the sphere, and A the angle which the arcs LL', $L''L'''$ make at their point of intersection, then we shall have*

$$\cos LL'' \,.\, \cos L'L''' - \cos LL''' \,.\, \cos L'L'' = \sin LL' \,.\, \sin L''L''' \,.\, \cos A$$

Demonstration. Let A denote also the point of intersection itself, and set

$$AL = t, \quad AL' = t', \quad AL'' = t'', \quad AL''' = t'''$$

Then we shall have

$$\cos LL'' = \cos t \,.\, \cos t'' + \sin t \sin t'' \cos A$$
$$\cos L'L''' = \cos t' \cos t''' + \sin t' \sin t''' \cos A$$
$$\cos LL''' = \cos t \cos t''' + \sin t \sin t''' \cos A$$
$$\cos L'L'' = \cos t' \cos t'' + \sin t' \sin t'' \cos A$$

and consequently,

$$\cos LL'' \,.\, \cos L'L''' - \cos LL''' \,.\, \cos L'L''$$
$$= \cos A \,(\cos t \cos t'' \sin t' \sin t''' + \cos t' \cos t''' \sin t \sin t''$$
$$- \cos t \cos t''' \sin t' \sin t'' - \cos t' \cos t'' \sin t \sin t''')$$
$$= \cos A\,(\cos t \sin t' - \sin t \cos t')\,(\cos t'' \sin t''' - \sin t'' \cos t''') \quad [5]$$
$$= \cos A \,.\, \sin(t' - t) \,.\, \sin(t''' - t'')$$
$$= \cos A \,.\, \sin LL' \,.\, \sin L''L'''$$

But as there are for each great circle two branches going out from the point A, these two branches form at this point two angles whose sum is $180°$. But our analysis shows that those branches are to be taken whose directions are in the sense from the point L to L', and from the point L'' to L'''; and since great circles intersect in two points, it is clear that either of the two points can be chosen arbitrarily. Also, instead of the angle A, we can take the arc between the poles of the great circles of which the arcs $L\,L'$, $L''\,L'''$ are parts. But it is evident that those poles are to be chosen which are similarly placed with respect to these arcs; that is to say, when we go from L to L' and from L'' to L''', both of the two poles are to be on the right, or both on the left.

VII. Let L, L', L'' be the three points on the sphere and set, for brevity,

$$\cos(1)L = x,\quad \cos(2)L = y,\quad \cos(3)L = z$$
$$\cos(1)L' = x',\quad \cos(2)L' = y',\quad \cos(3)L' = z'$$
$$\cos(1)L'' = x'',\quad \cos(2)L'' = y'',\quad \cos(3)L'' = z''$$

and also

$$x\,y'\,z'' + x'\,y''\,z + x''\,y\,z' - x\,y''\,z' - x'\,y\,z'' - x''\,y'\,z = \Delta$$

Let λ denote the pole of the great circle of which LL' is a part, this pole being the one that is placed in the same position with respect to this arc as the point (1) is with respect to the arc (2)(3). Then we shall have, by the preceding theorem,

$$y\,z' - y'\,z = \cos(1)\lambda \,.\, \sin(2)(3) \,.\, \sin LL',$$

or, because $(2)(3) = 90°$,

$$y\,z' - y'\,z = \cos(1)\lambda \,.\, \sin LL',$$

and similarly,

$$z\,x' - z'\,x = \cos(2)\lambda \,.\, \sin LL'$$
$$x\,y' - x'\,y = \cos(3)\lambda \,.\, \sin LL'$$

Multiplying these equations by x'', y'', z'' respectively, and adding, we obtain, by means of the second of the theorems deduced in V,

$$\Delta = \cos \lambda\, L'' \,.\, \sin LL'$$

Now there are three cases to be distinguished. *First*, when L'' lies on the great circle of which the arc LL' is a part, we shall have $\lambda L'' = 90°$, and consequently, $\Delta = 0$. If L'' does not lie on that great circle, the *second* case will be when L'' is on the same side as λ; the *third* case when they are on opposite sides. In the last two cases the points L, L', L'' will form a spherical triangle, and in the second case these points will lie in the same order as the points (1), (2), (3), and in the opposite order in the third case.

Denoting the angles of this triangle simply by L, L', L'' and the perpendicular drawn on the sphere from the point L'' to the side LL' by p, we shall have

$$\sin p = \sin L \,.\, \sin LL'' = \sin L' \,.\, \sin L'L'',$$

and

$$\lambda\, L'' = 90^\circ \mp p,$$

the upper sign being taken for the second case, the lower for the third. From this it follows that

$$\pm \Delta = \sin L \,.\, \sin LL' \,.\, \sin LL'' = \sin L' \,.\, \sin LL' \,.\, \sin L'L''$$
$$= \sin L'' \,.\, \sin LL'' \,.\, \sin L'L''$$

Moreover, it is evident that the first case can be regarded as contained in the second or third, and it is easily seen that the expression $\pm \Delta$ represents six times the volume of the pyramid formed by the points L, L', L'' and the centre of the sphere. Whence, finally, it is clear that the expression $\pm \frac{1}{6} \Delta$ expresses generally the volume of any pyramid contained between the origin of coordinates and the three points whose coordinates are z, y, z; x', y', z'; x'', y'', z''.

3.

A curved surface is said to possess continuous curvature at one of its points A, if the directions of all the straight lines drawn from A to points of the surface at an infinitely small distance from A are deflected infinitely little from one and the same plane passing through A.[6] This plane is said to *touch* the surface at the point A. If this condition is not satisfied for any point, the continuity of the curvature is here interrupted, as happens, for example, at the vertex of a cone. The following investigations will be restricted to such surfaces, or to such parts of surfaces, as have the continuity of their curvature nowhere interrupted. We shall only observe now that the methods used to determine the position of the tangent plane lose their meaning at singular points, in which the continuity of the curvature is interrupted, and must lead to indeterminate solutions.

4.

The orientation of the tangent plane is most conveniently studied by means of the direction of the straight line normal to the plane at the point A, which is also called the normal to the curved surface at the point A. We shall represent the direction of this normal by the point L on the auxiliary sphere, and we shall set

$$\cos(1)L = X, \quad \cos(2)L = Y, \quad \cos(3)L = Z;$$

and denote the coordinates of the point A by x, y, z. Also let $x + dx$, $y + dy$, $z + dz$ be the coordinates of another point A' on the curved surface; ds its distance from A, which is infinitely small; and finally, let λ be the point on the sphere representing the direction of the element AA'. Then we shall have

$$dx = ds \,.\, \cos(1)\lambda, \qquad dy = ds \,.\, \cos(2)\lambda, \qquad dz = ds \,.\, \cos(3)\lambda$$

and, since $\lambda\, L$ must be equal to $90°$,

$$X \cos(1)\lambda + Y \cos(2)\lambda + Z \cos(3)\lambda = 0$$

By combining these equations we obtain

$$X\, dx + Y\, dy + Z\, dz = 0.$$

There are two general methods for defining the nature of a curved surface. The *first* uses the equation between the coordinates x, y, z, which we may suppose reduced to the form $W = 0$, where W will be a function of the indeterminates x, y, z. Let the complete differential of the function W be

$$dW = P\, dx + Q\, dy + R\, dz$$

and on the curved surface we shall have

$$P\, dx + Q\, dy + R\, dz = 0$$

and consequently,

$$P \cos(1)\lambda + Q \cos(2)\lambda + R \cos(3)\lambda = 0$$

Since this equation, as well as the one we have established above, must be true for the directions of all elements ds on the curved surface, we easily see that X, Y, Z must be proportional to P, Q, R respectively, and consequently, since

$$X^2 + Y^2 + Z^2 = 1, \quad [7]$$

we shall have either

$$X = \frac{P}{\sqrt{(P^2 + Q^2 + R^2)}}, \qquad Y = \frac{Q}{\sqrt{(P^2 + Q^2 + R^2)}}, \qquad Z = \frac{R}{\sqrt{(P^2 + Q^2 + R^2)}}$$

or

$$X = \frac{-P}{\sqrt{(P^2 + Q^2 + R^2)}}, \qquad Y = \frac{-Q}{\sqrt{(P^2 + Q^2 + R^2)}}, \qquad Z = \frac{-R}{\sqrt{(P^2 + Q^2 + R^2)}}$$

The *second* method[8] expresses the coordinates in the form of functions of two variables, p, q. Suppose that differentiation of these functions gives

$$\begin{aligned} dx &= a\, dp + a'\, dq \\ dy &= b\, dp + b'\, dq \\ dz &= c\, dp + c'\, dq \end{aligned}$$

Substituting these values in the formula given above, we obtain

$$(a\,X + b\,Y + c\,Z)\,dp + (a'\,X + b'\,Y + c'\,Z)\,dq = 0$$

Since this equation must hold independently of the values of the differentials dp, dq, we evidently shall have

$$a\,X + b\,Y + c\,Z = 0, \quad a'\,X + b'\,Y + c'\,Z = 0$$

From this we see that X, Y, Z will be proportioned to the quantities

$$b\,c' - c\,b', \quad c\,a' - a\,c', \quad a\,b' - b\,a'$$

Hence, on setting, for brevity,

$$\sqrt{\left((b\,c' - c\,b')^2 + (c\,a' - a\,c')^2 + (a\,b' - b\,a')^2\right)} = \Delta$$

we shall have either

$$X = \frac{bc' - cb'}{\Delta}, \quad Y = \frac{ca' - ac'}{\Delta}, \quad Z = \frac{ab' - ba'}{\Delta}$$

or

$$X = \frac{cb' - bc'}{\Delta}, \quad Y = \frac{ac' - ca'}{\Delta}, \quad Z = \frac{ba' - ab'}{\Delta}$$

With these two general methods is associated a *third*, in which one of the coordinates, z, say, is expressed in the form of a function of the other two, x, y. This method is evidently only a particular case either of the first method, or of the second. If we set

$$dz = t\,dx + u\,dy$$

we shall have either

$$X = \frac{-t}{\sqrt{(1 + t^2 + u^2)}}, \quad Y = \frac{-u}{\sqrt{(1 + t^2 + u^2)}}, \quad Z = \frac{1}{\sqrt{(1 + t^2 + u^2)}}$$

or

$$X = \frac{t}{\sqrt{(1 + t^2 + u^2)}}, \quad Y = \frac{u}{\sqrt{(1 + t^2 + u^2)}}, \quad Z = \frac{-1}{\sqrt{(1 + t^2 + u^2)}}$$

5.

The two solutions found in the preceding article evidently refer to opposite points of the sphere, or to opposite directions, as one would expect, since the normal may be drawn toward either of the two sides of the curved surface. If we wish to distinguish between the two regions bordering upon the surface, and call one the exterior region and the other the interior region, we can then assign to each of the two normals its appropriate solution by aid of the theorem derived in Art. 2 (VII), and at the same time establish a criterion for distinguishing the one region from the other. [9]

In the first method, such a criterion is to be drawn from the sign of the quantity W. Indeed, generally speaking, the curved surface divides those regions of space in which W keeps a positive value from those in which the value of W becomes negative. In fact, it is easily seen from this theorem that, if W takes a positive value toward the exterior region, and if the normal is supposed to be drawn outwardly, the first solution is to be taken. Moreover, it will be easy to decide in any case whether the same rule for the sign of W is to hold throughout the entire surface, or whether for different parts there will be different rules. As long as the coefficients P, Q, R have finite values and do not all vanish at the same time, the law of continuity will prevent any change.

If we follow the second method, we can imagine two systems of curved lines on the curved surface, one system for which p is variable, q constant; the other for which q is variable, p constant. The respective positions of these lines with reference to the exterior region will decide which of the two solutions must be taken. In fact, whenever the three lines, namely, the branch of the line of the former system going out from the point A as p increases, the branch of the line of the latter system going out from the point A as q increases, and the normal drawn toward the exterior region, are *similarly* placed as the x, y, z axes respectively from the origin of abscissas (*e. g.*, if, both for the former three lines and for the latter three, we can conceive the first directed to the left, the second to the right, and the third upward), the first solution is to be taken. But whenever the relative position of the three lines is opposite to the relative position of the x, y, z axes, the second solution will hold.

In the third method, it is to be seen whether, when z receives a positive increment, x and y remaining constant, the point crosses toward the exterior or the interior region. In the former case, for the normal drawn outward, the first solution holds; in the latter case, the second.

6.

Just as each definite point on the curved surface is made to correspond to a definite point on the sphere, by the direction of the normal to the curved surface which is transferred to the surface of the sphere, so also any line whatever, or any figure whatever, on the latter will be represented by a corresponding line or figure on the former. In the comparison of two figures corresponding to one another in this way, one of which will be as the map of the other, two important points are to be considered, one when quantity alone is considered, the other when, disregarding quantitative relations, position alone is considered.

The first of these important points will be the basis of some ideas which it seems judicious to introduce into the theory of curved surfaces. Thus, to each part of a curved surface inclosed within definite limits we assign a *total* or *integral curvature*, which is represented by the area of the figure on the sphere corresponding to it. From this integral curvature must be distinguished the somewhat more specific curvature which we shall call the *measure of curvature*. The latter refers to a *point* of the surface, and shall denote the quotient obtained when the integral curvature of the surface element about a point is divided by the area of the element itself; and hence it denotes the ratio of the infinitely small areas which correspond to one another on the curved surface and on the sphere.[10] The use of these innovations will be abundantly justified, as we hope, by what we shall explain below. As for the terminology, we have thought it especially desirable that all ambiguity be avoided. For this reason we have not thought it advantageous to follow strictly the analogy of the terminology commonly adopted (though not approved by all) in the theory of plane curves, according to which the measure of curvature should be called simply curvature, but the total curvature, the amplitude. But why not be free in the choice of words, provided they are not meaningless and not liable to a misleading interpretation ?

The position of a figure on the sphere can be either similar to the position of the corresponding figure on the curved surface, or opposite (inverse). The former is the case when two lines going out on the curved surface from the same point in different, but not opposite directions, are represented on the sphere by lines similarly placed, that is, when the map of the line to the right is also to the right; the latter is the case when the contrary holds. We shall distinguish these two cases by the positive or negative *sign* of the measure of curvature. But evidently this distinction can hold only when on each surface we choose a definite face on which we suppose the figure to lie. On the auxiliary sphere we shall use always the exterior face, that is, that turned away from the centre; on the curved surface also there may be taken for the exterior face the one already considered, or rather that face from which the normal is supposed to be drawn. For, evidently, there is no change in regard to the similitude of the figures, if on the curved surface both the figure and the normal be transferred to the opposite side, so long as the image itself is represented on the same side of the sphere.

The positive or negative sign, which we assign to the *measure* of curvature according to the position of the infinitely small figure, we extend also to the integral curvature of a finite figure on the curved surface. However, if we wish to discuss the general case, some explanations will be necessary, which we can only touch here briefly. So long as the figure on the curved surface is such that to *distinct* points on itself there correspond distinct points on the sphere, the definition needs no further explanation. But whenever this condition is not satisfied, it will be necessary to take into account twice or several times certain parts of the figure on the sphere. Whence for a similar, or inverse position, may arise an accumulation of areas, or the areas may partially or wholly destroy each other. In such a case, the simplest way is to suppose the curved surface divided into parts, such that each part, considered separately, satisfies the above condition; to assign to each of the parts its integral curvature, determining this magnitude by the area of the corresponding figure on the sphere, and the sign by the position of this figure; and, finally, to assign to the total figure the integral curvature arising from the addition of the integral curvatures which correspond to the single parts. So, generally, the integral curvature of a figure is equal to $\int k\,d\sigma$, $d\sigma$ denoting the element of area of the figure, and k the measure of curvature at any point. The principal points concerning the geometric representation of this integral reduce to the following. To the perimeter of the figure on the curved surface (under the restriction of Art. 3) will correspond always a closed line on the sphere. If the latter nowhere intersect itself, it will divide the whole surface of the sphere into two parts, one of which will correspond to the figure on the curved surface; and its area (taken as positive or negative according as, with respect to its perimeter, its position is similar, or inverse, to the position of the figure on the curved surface) will represent the integral curvature of the figure on the curved surface. But whenever this line intersects itself once or several times, it will give a complicated figure, to which, however, it is possible to assign a definite area as legitimately as in the case of a figure without nodes; and this area, properly interpreted, will give always an exact value for the integral curvature. However, we must reserve for another occasion the more extended exposition of the theory of these figures viewed from this very general standpoint.[11]

7.

We shall now find a formula which will express the measure of curvature for any point of a curved surface. Let $d\sigma$ denote the area of an element of this surface; then $Z\,d\sigma$ will be the area of the projection of this element on the plane of the coordinates x, y; and consequently, if $d\Sigma$ is the area of the corresponding element on the sphere, $Z\,d\Sigma$ will be the area of its projection on the same plane. The positive or negative sign of Z will, in fact, indicate that the position of the projection is similar or inverse to that of the projected element. Evidently these projections have the same ratio as to quantity and the same relation as to position as the elements themselves. Let us consider now a triangular element on the curved surface,[12] and let us suppose that the coordinates of the three points which form its projection are

$$\begin{array}{ll} x, & y \\ x + dx, & y + dy \\ x + \delta x, & y + \delta y \end{array}$$

The double area of this triangle will be expressed by the formula

$$dx \,.\, \delta y - dy \,.\, \delta x$$

and this will be in a positive or negative form according as the position of the side from the first point to the third, with respect to the side from the first point to the second, is similar or opposite to the position of the y-axis of coordinates with respect to the x-axis of coordinates.

In like manner, if the coordinates of the three points which form the projection of the corresponding element on the sphere, from the centre of the sphere as origin, are

$$\begin{array}{ll} X, & Y \\ X+dX, & Y+dY \\ X+\delta X, & Y+\delta Y \end{array}$$

the double area of this projection will be expressed by

$$dX \,.\, \delta Y - dY \,.\, \delta X$$

and the sign of this expression is determined in the same manner as above. Wherefore the measure of curvature at this point of the curved surface will be

$$k = \frac{dX \,.\, \delta Y - dY \,.\, \delta X}{dx \,.\, \delta y - dy \,.\, \delta x}$$

If now we suppose the nature of the curved surface to be defined according to the third method considered in Art. 4, X and Y will be in the form of functions of the quantities x, y. We shall have, therefore, [13]

$$dX = \frac{\partial X}{\partial x} dx + \frac{\partial X}{\partial y} dy$$

$$\delta X = \frac{\partial X}{\partial x} \delta x + \frac{\partial X}{\partial y} \delta y$$

$$dY = \frac{\partial Y}{\partial x} dx + \frac{\partial Y}{\partial y} dy$$

$$\delta Y = \frac{\partial Y}{\partial x} \delta x + \frac{\partial Y}{\partial y} \delta y$$

When these values have been substituted, the above expression becomes

$$k = \frac{\partial X}{\partial x} \cdot \frac{\partial Y}{\partial y} - \frac{\partial X}{\partial y} \cdot \frac{\partial Y}{\partial x}$$

Setting, as above,

$$\frac{\partial z}{\partial x} = t, \quad \frac{\partial z}{\partial y} = u$$

and also

$$\frac{\partial^2 z}{\partial x^2} = T, \quad \frac{\partial^2 z}{\partial x \,.\, \partial y} = U, \quad \frac{\partial^2 z}{\partial y^2} = V$$

or

$$dt = T\,dx + U\,dy, \quad du = U\,dx + V\,dy$$

we have from the formulæ given above

$$X = -tZ, \quad Y = -uZ, \quad (1 + t^2 + u^2)\,Z^2 = 1$$

and hence

$$dX = -Z\,dt - t\,dZ$$
$$dY = -Z\,du - u\,dZ$$
$$(1 + t^2 + u^2)\,dZ + Z\,(t\,dt + u\,du) = 0$$

or

$$dZ = -Z^3\,(t\,dt + u\,du)$$
$$dX = -Z^3\,(1 + u^2)\,dt + Z^3\,tu\,du$$
$$dY = +Z^3\,tu\,dt - Z^3\,(1 + t^2)\,du$$ [14]

and so

$$\frac{\partial X}{\partial x} = Z^3\,(-(1 + u^2)\,T + tu\,U)$$
$$\frac{\partial X}{\partial y} = Z^3\,(-(1 + u^2)\,U + tu\,V)$$
$$\frac{\partial Y}{\partial x} = Z^3\,(tu\,T - (1 + t^2)\,U)$$
$$\frac{\partial Y}{\partial y} = Z^3\,(tu\,U - (1 + t^2)\,V)$$

Substituting these values in the above expression, it becomes

$$k = Z^6\,(TV - U^2)\,(1 + t^2 + u^2) = Z^4\,(TV - U^2)$$
$$= \frac{TV - U^2}{(1 + t^2 + u^2)^2}$$

8.

By a suitable choice of origin and axes of coordinates, we can easily make the values of the quantities t, u, U vanish for a definite point A. Indeed, the first two conditions will be fulfilled at once if the tangent plane at this point be taken for the xy-plane. If, further, the origin is placed at the point A itself, the expression for the coordinate z evidently takes the form

$$z = \tfrac{1}{2} T^\circ x^2 + U^\circ xy + \tfrac{1}{2} V^\circ y^2 + \Omega$$

where Ω will be of higher degree than the second. Turning now the axes of x and y through an angle M such that

$$\tan 2 M = \frac{2 U^\circ}{T^\circ - V^\circ}$$

it is easily seen that there must result an equation of the form

$$z = \tfrac{1}{2} Tx^2 + \tfrac{1}{2} Vy^2 + \Omega$$

In this way the third condition is also satisfied. When this has been done, it is evident that

I. If the curved surface be cut by a plane passing through the normal itself and through the x-axis, a plane curve will be obtained, the radius of curvature of which at the point A will be equal to $\frac{1}{T}$, the positive or negative sign indicating that the curve is concave or convex toward that region toward which the coordinates z are positive.

II. In like manner $\frac{1}{V}$ will be the radius of curvature at the point A of the plane curve which is the intersection of the surface and the plane through the y-axis and the z-axis.

III. Setting $x = r \cos \phi$, $y = r \sin \phi$, the equation becomes

$$z = \tfrac{1}{2} (T \cos^2 \phi + V \sin^2 \phi) r^2 + \Omega$$

from which we see that if the section is made by a plane through the normal at A and making an angle ϕ with the x-axis, we shall have a plane curve whose radius of curvature at the point A will be

$$\frac{1}{T \cos^2 \phi + V \sin^2 \phi}$$

IV. Therefore, whenever we have $T = V$, the radii of curvature in *all* the normal planes will be equal. But if T and V are not equal, it is evident that, since for any value whatever of the angle ϕ, $T \cos^2 \phi + V \sin^2 \phi$ falls between T and V, the radii of curvature in the principal sections considered in I. and II. refer to the extreme curvatures; that is to say, the one to the maximum curvature, the other to the minimum, if T and V have the same sign. On the other hand, one has the greatest convex curvature, the other the greatest concave curvature, if T and V have opposite signs. These conclusions contain almost all that the illustrious Euler was the first to prove on the curvature of curved surfaces. [15]

V. The measure of curvature at the point A on the curved surface takes the very simple form

$$k = TV,$$

whence we have the

THEOREM. *The measure of curvature at any point whatever of the surface is equal to a fraction whose numerator is unity, and whose denominator is the product of the two extreme radii of curvature of the sections by normal planes.*

At the same time it is clear that the measure of curvature is positive for concavo-concave or convexo-convex surfaces (which distinction is not essential), but negative for concavo-convex surfaces. If the surface consists of parts of each kind, then on the lines separating the two kinds the measure of curvature ought to vanish. Later we shall make a detailed study of the nature of curved surfaces for which the measure of curvature everywhere vanishes.

9.

The general formula for the measure of curvature given at the end of Art. 7 is the most simple of all, since it involves only five elements. We shall arrive at a more complicated formula, indeed, one involving nine elements, if we wish to use the first method of representing a curved surface. Keeping the notation of Art. 4, let us set also

$$\frac{\partial^2 W}{\partial x^2} = P', \quad \frac{\partial^2 W}{\partial y^2} = Q', \quad \frac{\partial^2 W}{\partial z^2} = R'$$

$$\frac{\partial^2 W}{\partial y . \partial z} = P'', \quad \frac{\partial^2 W}{\partial x . \partial z} = Q'', \quad \frac{\partial^2 W}{\partial x . \partial y} = R''$$

so that

$$dP = P'\, dx + R''\, dy + Q''\, dz$$
$$dQ = R''\, dx + Q'\, dy + P''\, dz$$
$$dR = Q''\, dx + P''\, dy + R'\, dz$$

Now since $t = -\frac{P}{R}$, we find through differentiation

$$R^2\, dt = -R\, dP + P\, dR = (PQ'' - RP')\, dx + (PP'' - RR'')\, dy + (PR' - RQ'')\, dz$$

or, eliminating dz by means of the equation

$$P\,dx + Q\,dy + R\,dz = 0,$$

$$R^3\,dt = (-R^2P' + 2PRQ'' - P^2R')\,dx + (PRP'' + QRQ'' - PQR' - R^2R'')\,dy.$$

In like manner we obtain

$$R^3\,du = (PRP'' + QRQ'' - PQR' - R^2R'')\,dx + (-R^2Q' + 2\,QRP'' - Q^2R')\,dy$$

From this we conclude that

$$R^3T = -R^2P' + 2\,PRQ'' - P^2R'$$
$$R^3U = PRP'' + QRQ'' - PQR' - R^2R''$$
$$R^3V = -R^2Q' + 2\,QRP'' - Q^2R'$$

Substituting these values in the formula of Art. 7, we obtain for the measure of curvature k the following symmetric expression:

$$(P^2 + Q^2 + R^2)^2 k = P^2(Q'R' - P''^2) + Q^2(P'R' - Q''^2) + R^2(P'Q' - R''^2)$$
$$+ 2\,QR(Q''R'' - P'P'') + 2\,PR(P''R'' - Q'Q'') + 2\,PQ(P''Q'' - R'R'')$$

10.

We obtain a still more complicated formula, indeed, one involving fifteen elements, if we follow the second general method of defining the nature of a curved surface. It is, however, very important that we develop this formula also. Retaining the notations of Art. 4, let us put also

$$\frac{\partial^2 x}{\partial p^2} = \alpha, \quad \frac{\partial^2 x}{\partial p\,.\,\partial q} = \alpha', \quad \frac{\partial^2 x}{\partial q^2} = \alpha''$$

$$\frac{\partial^2 y}{\partial p^2} = \beta, \quad \frac{\partial^2 y}{\partial p\,.\,\partial q} = \beta', \quad \frac{\partial^2 y}{\partial q^2} = \beta''$$

$$\frac{\partial^2 z}{\partial p^2} = \gamma, \quad \frac{\partial^2 z}{\partial p\,.\,\partial q} = \gamma', \quad \frac{\partial^2 z}{\partial q^2} = \gamma''$$

and let us put, for brevity,

$$bc' - cb' = A$$
$$ca' - ac' = B$$
$$ab' - ba' = C$$

First we see that

$$A\,dx + B\,dy + C\,dz = 0,$$

or

$$dz = -\frac{A}{C}dx - \frac{B}{C}dy.$$

Thus, inasmuch as z may be regarded as a function of x, y, we have

$$\frac{\partial z}{\partial x} = t = -\frac{A}{C}$$

$$\frac{\partial z}{\partial y} = u = -\frac{B}{C}$$

Then from the formulæ

$$dx = a\,dp + a'\,dq, \quad dy = b\,dp + b'\,dq,$$

we have

$$C\,dp = \quad b'\,dx - a'\,dy$$
$$C\,dq = -b\,dx + a\,dy$$

Thence we obtain for the total differentials of t, u

$$C^3\,dt = \left(A\frac{\partial C}{\partial p} - C\frac{\partial A}{\partial p}\right)(b'\,dx - a'\,dy) + \left(C\frac{\partial A}{\partial q} - A\frac{\partial C}{\partial q}\right)(b\,dx - a\,dy)$$

$$C^3\,du = \left(B\frac{\partial C}{\partial p} - C\frac{\partial B}{\partial p}\right)(b'\,dx - a'\,dy) + \left(C\frac{\partial B}{\partial q} - B\frac{\partial C}{\partial q}\right)(b\,dx - a\,dy)$$

If now we substitute in these formulæ

$$\frac{\partial A}{\partial p} = c'\,\beta + b\,\gamma' - c\,\beta' - b'\,\gamma$$

$$\frac{\partial A}{\partial q} = c'\,\beta' + b\,\gamma'' - c\,\beta'' - b'\,\gamma'$$

$$\frac{\partial B}{\partial p} = a'\,\gamma + c\,\alpha' - a\,\gamma' - c'\,\alpha$$

$$\frac{\partial B}{\partial q} = a'\,\gamma' + c\,\alpha'' - a\,\gamma'' - c'\,\alpha'$$

$$\frac{\partial C}{\partial p} = b'\,\alpha + a\,\beta' - b\,\alpha' - a'\,\beta$$

$$\frac{\partial C}{\partial q} = b'\,\alpha' + a\,\beta'' - b\,\alpha'' - a'\,\beta'$$

and if we note that the values of the differentials dt, du thus obtained must be equal, independently of the differentials dx, dy, to the quantities $T\,dx + U\,dy$, $U\,dx + V\,dy$ respectively, we shall find, after some sufficiently obvious transformations,

$$\begin{aligned} C^3\,T = {} & \alpha A b'^2 + \beta B b'^2 + \gamma C b'^2 \\ & - 2\,\alpha' A b b' - 2\,\beta' B b b' - 2\,\gamma' C b b' \\ & + \alpha'' A b^2 + \beta'' B b^2 + \gamma'' C b^2 \end{aligned}$$

$$\begin{aligned}C^3 U = & -\alpha A a'b' - \beta B a'b' - \gamma C a'b' \\ & + \alpha' A(ab' + ba') + \beta' B(ab' + ba') + \gamma' C(ab' + ba') \\ & - \alpha'' A ab - \beta'' B ab - \gamma'' C ab \\ C^3 V = & \alpha A a'^2 + \beta B a'^2 + \gamma C a'^2 \\ & - 2\alpha' A aa' - 2\beta' B aa' - 2\gamma' C aa' \\ & + \alpha'' A a^2 + \beta'' B a^2 + \gamma'' C a^2\end{aligned}$$

Hence, if we put, for the sake of brevity,

$$A\alpha + B\beta + C\gamma = D \quad \ldots \ldots \quad (1)$$
$$A\alpha' + B\beta' + C\gamma' = D' \quad \ldots \ldots \quad (2)$$
$$A\alpha'' + B\beta'' + C\gamma'' = D'' \quad \ldots \ldots \quad (3)$$ [16]

we shall have

$$\begin{aligned}C^3 T &= D b'^2 - 2 D' bb' + D'' b^2 \\ C^3 U &= -D a'b' + D'(ab' + ba') - D'' ab \\ C^3 V &= D a'^2 - 2 D' aa' + D'' a^2\end{aligned}$$

From this we find, after the reckoning has been carried out,

$$C^6(TV - U^2) = (DD'' - D'^2)(ab' - ba')^2 = (DD'' - D'^2)C^2$$

and therefore the formula for the measure of curvature

$$k = \frac{DD'' - D'^2}{(A^2 + B^2 + C^2)^2}$$

11.

By means of the formula just found we are going to establish another, which may be counted among the most productive theorems in the theory of curved surfaces. Let us introduce the following notation:

$$a^2 + b^2 + c^2 = E$$
$$aa' + bb' + cc' = F$$
$$a'^2 + b'^2 + c'^2 = G$$
$$a\alpha + b\beta + c\gamma = m \quad \ldots \ldots \quad (4)$$
$$a\alpha' + b\beta' + c\gamma' = m' \quad \ldots \ldots \quad (5)$$
$$a\alpha'' + b\beta'' + c\gamma'' = m'' \quad \ldots \ldots \quad (6)$$
$$a'\alpha + b'\beta + c'\gamma = n \quad \ldots \ldots \quad (7)$$
$$a'\alpha' + b'\beta' + c'\gamma' = n' \quad \ldots \ldots \quad (8)$$
$$a'\alpha'' + b'\beta'' + c'\gamma'' = n'' \quad \ldots \ldots \quad (9)$$
$$A^2 + B^2 + C^2 = EG - F^2 = \Delta$$

Let us eliminate from the equations 1, 4, 7 the quantities β, γ, which is done by multiplying them by $bc'-cb'$, $b'C-c'B$, $cB-bC$ respectively and adding. In this way we obtain

$$\begin{aligned}(A(bc'-cb')+a(b'C-c'B)+a'(cB-bC))\,\alpha \\ =D(bc'-cb')+m(b'C-c'B)+n(cB-bC)\end{aligned}$$

an equation which is easily transformed into

$$AD=\alpha\Delta+a(nF-mG)+a'(mF-nE)$$

Likewise the elimination of α, γ or α, β from the same equations gives

$$\begin{aligned}BD&=\beta\Delta+b(nF-mG)+b'(mF-nE)\\ CD&=\gamma\Delta+c(nF-mG)+c'(mF-nE)\end{aligned}$$

Multiplying these three equations by α'', β'', γ'' respectively and adding, we obtain

$$DD''=(\alpha\alpha''+\beta\beta''+\gamma\gamma'')\Delta+m''(nF-mG)+n''(mF-nE) \quad . \; . \; . \quad (10)$$

If we treat the equations 2, 5, 8 in the same way, we obtain

$$\begin{aligned}AD'&=\alpha'\Delta+a(n'F-m'G)+a'(m'F-n'E)\\ BD'&=\beta'\Delta+b(n'F-m'G)+b'(m'F-n'E)\\ CD'&=\gamma'\Delta+c(n'F-m'G)+c'(m'F-n'E)\end{aligned}$$

and after these equations are multiplied by α', β', γ' respectively, addition gives

$$D'^2=(\alpha'^2+\beta'^2+\gamma'^2)\Delta+m'(n'F-m'G)+n'(m'F-n'E)$$

A combination of this equation with equation (10) gives

$$\begin{aligned}DD''-D'^2=(\alpha\alpha''+\beta\beta''+\gamma\gamma''-\alpha'^2-\beta'^2-\gamma'^2)\Delta \\ +E(n'^2-nn'')+F(nm''-2m'n'+mn'')+G(m'^2-mm'')\end{aligned}$$

It is clear that we have [17]

$$\frac{\partial E}{\partial p}=2m,\quad \frac{\partial E}{\partial q}=2m',\quad \frac{\partial F}{\partial p}=m'+n,\quad \frac{\partial F}{\partial q}=m''+n',\quad \frac{\partial G}{\partial p}=2n',\quad \frac{\partial G}{\partial q}=2n'',$$

or

$$m=\tfrac{1}{2}\frac{\partial E}{\partial p},\qquad m'=\tfrac{1}{2}\frac{\partial E}{\partial q},\quad m''=\frac{\partial F}{\partial q}-\tfrac{1}{2}\frac{\partial G}{\partial p}$$

$$n=\frac{\partial F}{\partial p}-\tfrac{1}{2}\frac{\partial E}{\partial q},\quad n'=\tfrac{1}{2}\frac{\partial G}{\partial p},\quad n''=\tfrac{1}{2}\frac{\partial G}{\partial q}$$

Moreover, it is easily shown that we shall have

$$\begin{aligned}\alpha\alpha''+\beta\beta''+\gamma\gamma''-\alpha'^2-\beta'^2-\gamma'^2&=\frac{\partial n}{\partial q}-\frac{\partial n'}{\partial p}=\frac{\partial m''}{\partial p}-\frac{\partial m'}{\partial q}\\ &=-\tfrac{1}{2}\cdot\frac{\partial^2 E}{\partial q^2}+\frac{\partial^2 F}{\partial p.\partial q}-\tfrac{1}{2}\cdot\frac{\partial^2 G}{\partial p^2}\end{aligned}$$

If we substitute these different expressions in the formula for the measure of curvature derived at the end of the preceding article, we obtain the following formula, which involves only the quantities E, F, G and their differential quotients of the first and second orders:

$$4(EG-F^2)^2 k = E\left(\frac{\partial E}{\partial q}\cdot\frac{\partial G}{\partial q}-2\frac{\partial F}{\partial p}\cdot\frac{\partial G}{\partial q}+\left(\frac{\partial G}{\partial p}\right)^2\right)$$

$$+F\left(\frac{\partial E}{\partial p}\cdot\frac{\partial G}{\partial q}-\frac{\partial E}{\partial q}\cdot\frac{\partial G}{\partial p}-2\frac{\partial E}{\partial q}\cdot\frac{\partial F}{\partial q}+4\frac{\partial F}{\partial p}\cdot\frac{\partial F}{\partial q}-2\frac{\partial F}{\partial p}\cdot\frac{\partial G}{\partial p}\right)$$

$$+G\left(\frac{\partial E}{\partial p}\cdot\frac{\partial G}{\partial p}-2\frac{\partial E}{\partial p}\cdot\frac{\partial F}{\partial q}+\left(\frac{\partial E}{\partial q}\right)^2\right)-2(EG-F^2)\left(\frac{\partial^2 E}{\partial q^2}-2\frac{\partial^2 F}{\partial p\,.\,\partial q}+\frac{\partial^2 G}{\partial p^2}\right)$$

12.

Since we always have

$$dx^2+dy^2+dz^2 = E\,dp^2+2F\,dp\,.\,dq+G\,dq^2,$$

it is clear that

$$\sqrt{(E\,dp^2+2F\,dp\,.\,dq+G\,dq^2)}$$

is the general expression for the linear element on the curved surface. The analysis developed in the preceding article thus shows us that for finding the measure of curvature there is no need of finite formulæ, which express the coordinates x, y, z as functions of the indeterminates p, q; but that the general expression for the magnitude of any linear element is sufficient. Let us proceed to some applications of this very important theorem.

Suppose that our surface can be developed upon another surface, curved or plane, so that to each point of the former surface, determined by the coordinates x, y, z, will correspond a definite point of the latter surface, whose coordinates are x', y', z'. Evidently x', y', z' can also be regarded as functions of the indeterminates p, q, and therefore for the element $\sqrt{(dx'^2+dy'^2+dz'^2)}$ we shall have an expression of the form

$$\sqrt{(E'\,dp^2+2F'\,dp\,.\,dq+G'\,dq^2)}$$

where E', F', G' also denote functions of p, q. But from the very notion of the *development* of one surface upon another it is clear that the elements corresponding to one another on the two surfaces are necessarily equal. Therefore we shall have identically

$$E=E',\quad F=F',\quad G=G'.$$

Thus the formula of the preceding article leads of itself to the remarkable

THEOREM. *If a curved surface is developed upon any other surface whatever, the measure of curvature in each point remains unchanged.*

Also it is evident that *any finite part whatever of the curved surface will retain the same integral curvature after development upon another surface.*

Surfaces developable upon a plane constitute the particular case to which geometers have heretofore restricted their attention. Our theory shows at once that the measure of curvature at every point of such surfaces is equal to zero. Consequently, if the nature of these surfaces is defined according to the third method, we shall have at every point

$$\frac{\partial^2 z}{\partial x^2}\cdot\frac{\partial^2 z}{\partial y^2}-\left(\frac{\partial^2 z}{\partial x\,.\,\partial y}\right)^2=0$$

a criterion which, though indeed known a short time ago, has not, at least to our knowledge, commonly been demonstrated with as much rigor as is desirable.

13.

What we have explained in the preceding article is connected with a particular method of studying surfaces, a very worthy method which may be thoroughly developed by geometers. When a surface is regarded, not as the boundary of a solid, but as a flexible, though not extensible solid, one dimension of which is supposed to vanish, then the properties of the surface depend in part upon the form to which we can suppose it reduced, and in part are absolute and remain invariable, whatever may be the form into which the surface is bent. To these latter properties, the study of which opens to geometry a new and fertile field, belong the measure of curvature and the integral curvature, in the sense which we have given to these expressions. To these belong also the theory of shortest lines, and a great part of what we reserve to be treated later. From this point of view, a plane surface and a surface developable on a plane, *e. g.*, cylindrical surfaces, conical surfaces, etc., are to be regarded as essentially identical; and the generic method of defining in a general manner the nature of the surfaces thus considered is always based upon the formula

$$\sqrt{(E\,dp^2+2\,F\,dp\,.\,dq+G\,dq^2)},$$

which connects the linear element with the two indeterminates p, q.[18] But before following this study further, we must introduce the principles of the theory of shortest lines on a given curved surface.

14.

The nature of a curved line in space is generally given in such a way that the coordinates x, y, z corresponding to the different points of it are given in the form of functions of a single variable, which we shall call w. The length of such a line from an arbitrary initial point to the point whose coordinates are x, y, z, is expressed by the integral

$$\int dw \,.\sqrt{\left(\left(\frac{dx}{dw}\right)^2+\left(\frac{dy}{dw}\right)^2+\left(\frac{dz}{dw}\right)^2\right)}$$

If we suppose that the position of the line undergoes an infinitely small variation, so that the coordinates of the different points receive the variations δx, δy, δz, the variation of the whole length becomes

$$\int \frac{dx\,.\,d\,\delta x + dy\,.\,d\,\delta y + dz\,.\,d\,\delta z}{\sqrt{(dx^2+dy^2+dz^2)}}$$

which expression we can change into the form[19]

$$\frac{dx\,.\,\delta x + dy\,.\,\delta y + dz\,.\,\delta z}{\sqrt{(dx^2+dy^2+dz^2)}}$$

$$-\int\left(\delta x\,.\,d\frac{dx}{\sqrt{(dx^2+dy^2+dz^2)}}+\delta y\,.\,d\frac{dy}{\sqrt{(dx^2+dy^2+dz^2)}}+\delta z\,.\,d\frac{dz}{\sqrt{(dx^2+dy^2+dz^2)}}\right)$$

We know that, in case the line is to be the shortest between its end points, all that stands under the integral sign must vanish. Since the line must lie on the given surface, whose nature is defined by the equation

$$P\,dx + Q\,dy + R\,dz = 0,$$

the variations δx, δy, δz also must satisfy the equation

$$P\,\delta x + Q\,\delta y + R\,\delta z = 0,$$

and from this it follows at once, according to well-known rules, that the differentials

$$d\frac{dx}{\sqrt{(dx^2+dy^2+dz^2)}},\quad d\frac{dy}{\sqrt{(dx^2+dy^2+dz^2)}},\quad d\frac{dz}{\sqrt{(dx^2+dy^2+dz^2)}}$$

must be proportional to the quantities P, Q, R respectively. Let dr be the element of the curved line; λ the point on the sphere representing the direction of this element; L the point on the sphere representing the direction of the normal to the curved surface; finally, let ξ, η, ζ be the coordinates of the point λ, and X, Y, Z be those of the point L with reference to the centre of the sphere. We shall then have

$$dx = \xi\,dr,\quad dy = \eta\,dr,\quad dz = \zeta\,dr$$

from which we see that the above differentials become $d\xi$, $d\eta$, $d\zeta$. And since the quantities P, Q, R are proportional to X, Y, Z, the character of shortest lines is expressed by the equations

$$\frac{d\xi}{X} = \frac{d\eta}{Y} = \frac{d\zeta}{Z}$$

Moreover, it is easily seen that

$$\sqrt{(d\xi^2+d\eta^2+d\zeta^2)}$$

is equal to the small arc on the sphere which measures the angle between the directions of the tangents at the beginning and at the end of the element dr, and is thus equal to $\frac{dr}{\rho}$, if ρ denotes the radius of curvature of the shortest line at this point. Thus we shall have

$$\rho\, d\xi = X\,dr,\quad \rho\, d\eta = Y\,dr,\quad \rho\, d\zeta = Z\,dr$$

15.

Suppose that an infinite number of shortest lines go out from a given point A on the curved surface, and suppose that we distinguish these lines from one another by the angle that the first element of each of them makes with the first element of one of them which we take for the first. Let ϕ be that angle, or, more generally, a function of that angle, and r the length of such a shortest line from the point A to the point whose coordinates are x, y, z. Since to definite values of the variables r, ϕ there correspond definite points of the surface, the coordinates x, y, z can be regarded as functions of r, ϕ. We shall retain for the notation $\lambda, L, \xi, \eta, \zeta, X, Y, Z$ the same meaning as in the preceding article, this notation referring to any point whatever on any one of the shortest lines.

All the shortest lines that are of the same length r will end on another line whose length, measured from an arbitrary initial point, we shall denote by v. Thus v can be regarded as a function of the indeterminates r, ϕ, and if λ' denotes the point on the sphere corresponding to the direction of the element dv, and also ξ', η', ζ' denote the coordinates of this point with reference to the centre of the sphere, we shall have

$$\frac{\partial x}{\partial \phi} = \xi' \cdot \frac{\partial v}{\partial \phi},\quad \frac{\partial y}{\partial \phi} = \eta' \cdot \frac{\partial v}{\partial \phi},\quad \frac{\partial z}{\partial \phi} = \zeta' \cdot \frac{\partial v}{\partial \phi}$$

From these equations and from the equations

$$\frac{\partial x}{\partial r} = \xi,\quad \frac{\partial y}{\partial r} = \eta,\quad \frac{\partial z}{\partial r} = \zeta$$

we have

$$\frac{\partial x}{\partial r} \cdot \frac{\partial x}{\partial \phi} + \frac{\partial y}{\partial r} \cdot \frac{\partial y}{\partial \phi} + \frac{\partial z}{\partial r} \cdot \frac{\partial z}{\partial \phi} = (\xi\xi' + \eta\eta' + \zeta\zeta') \cdot \frac{\partial v}{\partial \phi} = \cos \lambda\lambda' \cdot \frac{\partial v}{\partial \phi}.$$

Let S denote the first member of this equation, which will also be a function of r, ϕ. Differentiation of S with respect to r gives

$$\frac{\partial S}{\partial r}=\frac{\partial^2 x}{\partial r^2}\cdot\frac{\partial x}{\partial \phi}+\frac{\partial^2 y}{\partial r^2}\cdot\frac{\partial y}{\partial \phi}+\frac{\partial^2 z}{\partial r^2}\cdot\frac{\partial z}{\partial \phi}+\tfrac{1}{2}\cdot\frac{\partial\left(\left(\frac{\partial x}{\partial r}\right)^2+\left(\frac{\partial y}{\partial r}\right)^2+\left(\frac{\partial z}{\partial r}\right)^2\right)}{\partial \phi}$$

$$=\frac{\partial \xi}{\partial r}\cdot\frac{\partial x}{\partial \phi}+\frac{\partial \eta}{\partial r}\cdot\frac{\partial y}{\partial \phi}+\frac{\partial \zeta}{\partial r}\cdot\frac{\partial z}{\partial \phi}+\tfrac{1}{2}\cdot\frac{\partial(\xi^2+\eta^2+\zeta^2)}{\partial \phi}$$

But

$$\xi^2+\eta^2+\zeta^2=1,$$

and therefore its differential is equal to zero; and by the preceding article we have, if ρ denotes the radius of curvature of the line r,

$$\frac{\partial \xi}{\partial r}=\frac{X}{\rho},\quad \frac{\partial \eta}{\partial r}=\frac{Y}{\rho},\quad \frac{\partial \zeta}{\partial r}=\frac{Z}{\rho}$$

Thus we have

$$\frac{\partial S}{\partial r}=\frac{1}{\rho}\cdot(X\xi'+Y\eta'+Z\zeta')\cdot\frac{\partial v}{\partial \phi}=\frac{1}{\rho}\cdot\cos L\lambda'\cdot\frac{\partial v}{\partial \phi}=0$$

since λ' evidently lies on the great circle whose pole is L. From this we see that S is independent of r, and is, therefore, a function of ϕ alone. But for $r=0$ we evidently have $v=0$, consequently $\frac{\partial v}{\partial \phi}=0$, and $S=0$ independently of ϕ. Thus, in general, we have necessarily $S=0$, and so $\cos\lambda\lambda'=0$, *i. e.*, $\lambda\lambda'=90°$. From this follows the

THEOREM. *If on a curved surface an infinite number of shortest lines of equal length be drawn from the same initial point, the lines joining their extremities will be normal to each of the lines.*

We have thought it worth while to deduce this theorem from the fundamental property of shortest lines; but the truth of the theorem can be made apparent without any calculation by means of the following reasoning. Let AB, AB' be two shortest lines of the same length including at A an infinitely small angle, and let us suppose that one of the angles made by the element BB' with the lines BA, $B'A$ differs from a right angle by a finite quantity. Then, by the law of continuity, one will be greater and the other less than a right angle. Suppose the angle at B is equal to $90°-\omega$, and take on the line AB a point C, such that

$$BC=BB'.\operatorname{cosec}\omega.$$

Then, since the infinitely small triangle $BB'C$ may be regarded as plane, we shall have

$$CB'=BC.\cos\omega,$$

and consequently

$$AC + CB' = AC + BC \,.\, \cos\omega = AB - BC \,.\, (1 - \cos\omega) = AB' - BC \,.\, (1 - \cos\omega),$$

i. e., the path from A to B' through the point C is shorter than the shortest line, *Q. E. A.*

16.

With the theorem of the preceding article we associate another, which we state as follows: *If on a curved surface we imagine any line whatever, from the different points of which are drawn at right angles and toward the same side an infinite number of shortest lines of the same length, the curve which joins their other extremities will cut each of the lines at right angles.* For the demonstration of this theorem no change need be made in the preceding analysis, except that ϕ must denote the length of the *given* curve measured from an arbitrary point; or rather, a function of this length. Thus all of the reasoning will hold here also, with this modification, that $S=0$ for $r=0$ is now implied in the hypothesis itself. Moreover, this theorem is more general than the preceding one, for we can regard it as including the first one if we take for the given line the infinitely small circle described about the centre A. Finally, we may say that here also geometric considerations may take the place of the analysis, which, however, we shall not take the time to consider here, since they are sufficiently obvious.

17.

We return to the formula

$$\sqrt{(E\,dp^2 + 2\,F\,dp\,.\,dq + G\,dq^2)},$$

which expresses generally the magnitude of a linear element on the curved surface, and investigate, first of all, the geometric meaning of the coefficients E, F, G. We have already said in Art. 5 that two systems of lines may be supposed to lie on the curved surface, p being variable, q constant along each of the lines of the one system; and q variable, p constant along each of the lines of the other system. Any point whatever on the surface can be regarded as the intersection of a line of the first system with a line of the second; and then the element of the first line adjacent to this point and corresponding to a variation dp will be equal to $\sqrt{E}\,.\,dp$, and the element of the second line corresponding to the variation dq will be equal to $\sqrt{G}\,.\,dq$. Finally, denoting by ω the angle between these elements, it is easily seen that we shall have

$$\cos\omega = \frac{F}{\sqrt{EG}}. \qquad 20$$

Furthermore, the area of the surface element in the form of a parallelogram between the two lines of the first system, to which correspond q, $q+dq$, and the two lines of the second system, to which correspond p, $p+dp$, will be

$$\sqrt{(EG-F^2)}\, dp\,.\,dq.$$

Any line whatever on the curved surface belonging to neither of the two systems is determined when p and q are supposed to be functions of a new variable, or one of them is supposed to be a function of the other. Let s be the length of such a curve, measured from an arbitrary initial point, and in either direction chosen as positive. Let θ denote the angle which the element

$$ds=\sqrt{(E\,dp^2+2\,F\,dp\,.\,dq+G\,dq^2)}$$

makes with the line of the first system drawn through the initial point of the element, and, in order that no ambiguity may arise, let us suppose that this angle is measured from that branch of the first line on which the values of p increase, and is taken as positive toward that side toward which the values of q increase. These conventions being made, it is easily seen that

$$\cos\theta\,.\,ds=\sqrt{E}\,.\,dp+\sqrt{G}\,.\cos\omega\,.\,dq=\frac{E\,dp+F\,dq}{\sqrt{E}}$$

$$\sin\theta\,.\,ds=\sqrt{G}\,.\sin\omega\,.\,dq=\frac{\sqrt{(EG-F^2)}\,.\,dq}{\sqrt{E}}$$

18.

We shall now investigate the condition that this line be a shortest line. Since its length s is expressed by the integral

$$s=\int\sqrt{(E\,dp^2+2\,F\,dp\,.\,dq+G\,dq^2)}$$

the condition for a minimum requires that the variation of this integral arising from an infinitely small change in the position become equal to zero. The calculation, for our purpose, is more simply made in this case, if we regard p as a function of q. When this is done, if the variation is denoted by the characteristic δ, we have

$$\delta s=\int\frac{\left(\frac{\partial E}{\partial p}\cdot dp^2+2\frac{\partial F}{\partial p}\cdot dp\,.\,dq+\frac{\partial G}{\partial p}\cdot dq^2\right)\delta p+(2\,E\,dp+2\,F\,dq)\,d\,\delta p}{2\,ds}$$

$$=\frac{E\,dp+F\,dq}{ds}\cdot\delta p+$$

$$+\int \delta p\left(\frac{\frac{\partial E}{\partial p}\cdot dp^2+2\frac{\partial F}{\partial p}\cdot dp\,.\,dq+\frac{\partial G}{\partial p}\cdot dq^2}{2\,ds}-d\,.\,\frac{E\,dp+F\,dq}{ds}\right)$$

and we know that what is included under the integral sign must vanish independently of δp. Thus we have

$$\begin{aligned}\frac{\partial E}{\partial p}\cdot dp^2+2\frac{\partial F}{\partial p}\cdot dp\,.\,dq+\frac{\partial G}{\partial p}\cdot dq^2&=2\,ds\,.\,d\,.\,\frac{E\,dp+F\,dq}{ds}\\ &=2\,ds\,.\,d\,.\sqrt{E}\,.\cos\theta\\ &=\frac{ds\,.\,dE\,.\cos\theta}{\sqrt{E}}-2\,ds\,.\,d\theta\,.\sqrt{E}\,.\sin\theta\\ &=\frac{(E\,dp+F\,dq)\,dE}{E}-2\sqrt{(EG-F^2)}\,.\,dq\,.\,d\theta.\\ &=\left(\frac{E\,dp+F\,dq}{E}\right)\cdot\left(\frac{\partial E}{\partial p}\cdot dp+\frac{\partial E}{\partial q}\cdot dq\right)-2\sqrt{(EG-F^2)}\,.\,dq\,.\,d\theta\end{aligned}$$

This gives the following conditional equation for a shortest line:

$$\begin{aligned}\sqrt{(EG-F^2)}\,.\,d\theta=\frac{1}{2}\cdot\frac{F}{E}\cdot\frac{\partial E}{\partial p}\cdot dp+\frac{1}{2}\cdot\frac{F}{E}\cdot\frac{\partial E}{\partial q}\cdot dq+\frac{1}{2}\cdot\frac{\partial E}{\partial q}\cdot dp\\ -\frac{\partial F}{\partial p}\cdot dp-\frac{1}{2}\cdot\frac{\partial G}{\partial p}\cdot dq\end{aligned}$$

which can also be written

$$\sqrt{(EG-F^2)}\,.\,d\theta=\frac{1}{2}\cdot\frac{F}{E}\cdot dE+\frac{1}{2}\cdot\frac{\partial E}{\partial q}\cdot dp-\frac{\partial F}{\partial p}\cdot dp-\frac{1}{2}\cdot\frac{\partial G}{\partial p}\cdot dq$$

From this equation, by means of the equation

$$\cot\theta=\frac{E}{\sqrt{(EG-F^2)}}\cdot\frac{dp}{dq}+\frac{F}{\sqrt{(EG-F^2)}}$$

it is also possible to eliminate the angle θ, and to derive a differential equation of the second order between p and q, which, however, would become more complicated and less useful for applications than the preceding.

19.

The general formulæ, which we have derived in Arts. 11, 18 for the measure of curvature and the variation in the direction of a shortest line, become much simpler if the quantities p, q are so chosen that the lines of the first system cut everywhere orthogonally the lines of the second system; *i. e.*, in such a way that we have generally $\omega = 90°$, or $F = 0$. Then the formula for the measure of curvature becomes

$$4\,E^2 G^2 k = E \,.\, \frac{\partial E}{\partial q} \cdot \frac{\partial G}{\partial q} + E\left(\frac{\partial G}{\partial p}\right)^2 + G \cdot \frac{\partial E}{\partial p} \cdot \frac{\partial G}{\partial p} + G\left(\frac{\partial E}{\partial q}\right)^2 - 2\,EG\left(\frac{\partial^2 E}{\partial q^2} + \frac{\partial^2 G}{\partial p^2}\right),$$

and for the variation of the angle θ

$$\sqrt{EG} \,.\, d\theta = \frac{1}{2} \cdot \frac{\partial E}{\partial q} \cdot dp - \frac{1}{2} \cdot \frac{\partial G}{\partial p} \cdot dq$$

Among the various cases in which we have this condition of orthogonality, the most important is that in which all the lines of one of the two systems, *e. g.*, the first, are shortest lines. Here for a constant value of q the angle θ becomes equal to zero, and therefore the equation for the variation of θ just given shows that we must have $\frac{\partial E}{\partial q} = 0$, or that the coefficient E must be independent of q; *i. e.*, E must be either a constant or a function of p alone. It will be simplest to take for p the length of each line of the first system, which length, when all the lines of the first system meet in a point, is to be measured from this point, or, if there is no common intersection, from any line whatever of the second system. Having made these conventions, it is evident that p and q denote now the same quantities that were expressed in Arts. 15, 16 by r and ϕ, and that $E = 1$. Thus the two preceding formulæ become:

$$4\,G^2\,k = \left(\frac{\partial G}{\partial p}\right)^2 - 2\,G\frac{\partial^2 G}{\partial p^2}$$

$$\sqrt{G} \,.\, d\theta = -\frac{1}{2} \cdot \frac{\partial G}{\partial p} \cdot dq$$

or, setting $\sqrt{G} = m$,

$$k = -\frac{1}{m} \cdot \frac{\partial^2 m}{\partial p^2}, \quad d\theta = -\frac{\partial m}{\partial p} \cdot dq$$

Generally speaking, m will be a function of p, q, and $m\,dq$ the expression for the element of any line whatever of the second system. But in the particular case where all the lines p go out from the same point, evidently we must have $m = 0$ for $p = 0$. Furthermore, in the case under discussion we will take for q the angle itself which the first element of any line whatever of the first system makes with the element of any one of the lines chosen arbitrarily. Then, since for an infinitely small value of p the element of a line of the second system (which can be regarded as a circle described with radius p) is equal to $p\,dq$, we shall have for an infinitely small value of p, $m = p$, and consequently, for $p = 0$, $m = 0$ at the same time, and $\frac{\partial m}{\partial p} = 1$.

20.

We pause to investigate the case in which we suppose that p denotes in a general manner the length of the shortest line drawn from a fixed point A to any other point whatever of the surface, and q the angle that the first element of this line makes with the first element of another given shortest line going out from A. Let B be a definite point in the latter line, for which $q=0$, and C another definite point of the surface, at which we denote the value of q simply by A. Let us suppose the points B, C joined by a shortest line, the parts of which, measured from B, we denote in a general way, as in Art. 18, by s; and, as in the same article, let us denote by θ the angle which any element ds makes with the element dp; finally, let us denote by θ°, θ' the values of the angle θ at the points B, C. We have thus on the curved surface a triangle formed by shortest lines. The angles of this triangle at B and C we shall denote simply by the same letters, and B will be equal to $180^{\circ}-\theta$, C to θ' itself. But, since it is easily seen from our analysis that all the angles are supposed to be expressed, not in degrees, but by numbers, in such a way that the angle $57^{\circ}17'45''$, to which corresponds an arc equal to the radius, is taken for the unit, we must set

$$\theta^{\circ}=\pi-B, \quad \theta'=C$$

where 2π denotes the circumference of the sphere. Let us now examine the integral curvature of this triangle, which is equal to

$$\int k\, d\sigma,$$

$d\sigma$ denoting a surface element of the triangle. Wherefore, since this element is expressed by $m\, dp\,.\,dq$, we must extend the integral

$$\iint m\, dp\,.\,dq$$

over the whole surface of the triangle. Let us begin by integration with respect to p, which, because

$$k=-\frac{1}{m}\cdot\frac{\partial^2 m}{\partial p^2},$$

gives

$$dq\,.\left(\text{const.}-\frac{\partial m}{\partial p}\right),$$

for the integral curvature of the area lying between the lines of the first system, to which correspond the values q, $q+dq$ of the second indeterminate. Since this integral curvature must vanish for $p = 0$, the constant introduced by integration must be equal to the value of $\frac{\partial m}{\partial q}$ for $p = 0$, *i. e.*, equal to unity. Thus we have

$$dq\left(1 - \frac{\partial m}{\partial p}\right),$$

where for $\frac{\partial m}{\partial p}$ must be taken the value corresponding to the end of this area on the line CB. But on this line we have, by the preceding article,

$$\frac{\partial m}{\partial q} \cdot dq = -d\theta,$$

whence our expression is changed into $dq + d\theta$. Now by a second integration, taken from $q = 0$ to $q = A$, we obtain for the integral curvature

$$A + \theta' - \theta^{\circ},$$

or

$$A + B + C - \pi.$$

The integral curvature is equal to the area of that part of the sphere which corresponds to the triangle, taken with the positive or negative sign according as the curved surface on which the triangle lies is concavo-concave or concavo-convex. For unit area will be taken the square whose side is equal to unity (the radius of the sphere), and then the whole surface of the sphere becomes equal to 4π. Thus the part of the surface of the sphere corresponding to the triangle is to the whole surface of the sphere as $\pm\ (A + B + C - \pi)$ is to 4π. This theorem, which, if we mistake not, ought to be counted among the most elegant in the theory of curved surfaces, may also be stated as follows:

The excess over 180° of the sum of the angles of a triangle formed by shortest lines on a concavo-concave curved surface, or the deficit from 180° of the sum of the angles of a triangle formed by shortest lines on a concavo-convex curved surface, is measured by the area of the part of the sphere which corresponds, through the directions of the normals, to that triangle, if the whole surface of the sphere is set equal to 720 degrees.

More generally, in any polygon whatever of n sides, each formed by a shortest line, the excess of the sum of the angles over $(2n - 4)$ right angles, or the deficit from $(2n - 4)$ right angles (according to the nature of the curved surface), is equal to the area of the corresponding polygon on the sphere, if the whole surface of the sphere is set equal to 720 degrees. This follows at once from the preceding theorem by dividing the polygon into triangles.

21.

Let us again give to the symbols p, q, E, F, G, ω the general meanings which were given to them above, and let us further suppose that the nature of the curved surface is defined in a similar way by two other variables, p', q', in which case the general linear element is expressed by

$$\sqrt{(E'\,dp'^2 + 2\,F'\,dp'.\,dq' + G'\,dq'^2)}$$

Thus to any point whatever lying on the surface and defined by definite values of the variables p, q will correspond definite values of the variables p', q', which will therefore be functions of p, q. Let us suppose we obtain by differentiating them

$$dp' = \alpha\,dp + \beta\,dq$$
$$dq' = \gamma\,dp + \delta\,dq$$

We shall now investigate the geometric meaning of the coefficients α, β, γ, δ.

Now *four* systems of lines may thus be supposed to lie upon the curved surface, for which p, q, p', q' respectively are constants. If through the definite point to which correspond the values p, q, p', q' of the variables we suppose the four lines belonging to these different systems to be drawn, the elements of these lines, corresponding to the positive increments dp, dq, dp', dq', will be

$$\sqrt{E}\,.\,dp, \quad \sqrt{G}\,.\,dq, \quad \sqrt{E'}\,.\,dp', \quad \sqrt{G'}\,.\,dq'.$$

The angles which the directions of these elements make with an arbitrary fixed direction we shall denote by M, N, M', N', measuring them in the sense in which the second is placed with respect to the first, so that $\sin(N-M)$ is positive. Let us suppose (which is permissible) that the fourth is placed in the same sense with respect to the third, so that $\sin(N'-M')$ also is positive. Having made these conventions, if we consider another point at an infinitely small distance from the first point, and to which correspond the values $p+dp$, $q+dq$, $p'+dp'$, $q'+dq'$ of the variables, we see without much difficulty that we shall have generally, *i. e.*, independently of the values of the increments dp, dq, dp', dq',

$$\sqrt{E}\,.\,dp\,.\,\sin M + \sqrt{G}\,.\,dq\,.\,\sin N = \sqrt{E'}\,.\,dp'\,.\,\sin M' + \sqrt{G'}\,.\,dq'\,.\,\sin N'$$

since each of these expressions is merely the distance of the new point from the line from which the angles of the directions begin. But we have, by the notation introduced above,

$$N - M = \omega.$$

In like manner we set

$$N' - M' = \omega',$$

and also

$$N - M' = \psi.$$

Then the equation just found can be thrown into the following form:

$$\sqrt{E}\,.\,dp\,.\sin(M' - \omega + \psi) + \sqrt{G}\,.\,dq\,.\sin(M' + \psi)$$
$$= \sqrt{E'}\,.\,dp'\,.\sin M' + \sqrt{G'}\,.\,dq'\,.\sin(M' + \omega')$$

or

$$\sqrt{E}\,.\,dp\,.\sin(N' - \omega - \omega' + \psi) + \sqrt{G}\,.\,dq\,.\sin(N' - \omega' + \psi)$$
$$= \sqrt{E'}\,.\,dp'\,.\sin(N' - \omega') + \sqrt{G'}\,.\,dq'\,.\sin N'$$

And since the equation evidently must be independent of the initial direction, this direction can be chosen arbitrarily. Then, setting in the second formula $N' = 0$, or in the first $M' = 0$, we obtain the following equations:

$$\sqrt{E'}\,.\sin\omega'\,.\,dp' = \sqrt{E}\,.\sin(\omega + \omega' - \psi)\,.\,dp + \sqrt{G}\,.\sin(\omega' - \psi)\,.\,dq$$
$$\sqrt{G'}\,.\sin\omega'\,.\,dq' = \sqrt{E}\,.\sin(\psi - \omega)\,.\,dp + \sqrt{G}\,.\sin\psi\,.\,dq$$

and these equations, since they must be identical with

$$dp' = \alpha\,dp + \beta\,dq$$
$$dq' = \gamma\,dp + \delta\,dq$$

determine the coefficients α, β, γ, δ. We shall have

$$\alpha = \sqrt{\frac{E}{E'}}\cdot\frac{\sin(\omega + \omega' - \psi)}{\sin\omega'},\quad \beta = \sqrt{\frac{G}{E'}}\cdot\frac{\sin(\omega' - \psi)}{\sin\omega'}$$
$$\gamma = \sqrt{\frac{E}{G'}}\cdot\frac{\sin(\psi - \omega)}{\sin\omega'},\quad \delta = \sqrt{\frac{G}{G'}}\cdot\frac{\sin\psi}{\sin\omega'}$$

These four equations, taken in connection with the equations

$$\cos\omega = \frac{F}{\sqrt{EG}},\qquad \cos\omega' = \frac{F'}{\sqrt{E'G'}},$$
$$\sin\omega = \sqrt{\frac{EG - F^2}{EG}},\quad \sin\omega' = \sqrt{\frac{E'G' - F'^2}{E'G'}},$$

may be written

$$\alpha\sqrt{(E'G' - F'^2)} = \sqrt{EG'}\,.\sin(\omega + \omega' - \psi)$$
$$\beta\sqrt{(E'G' - F'^2)} = \sqrt{GG'}\,.\sin(\omega' - \psi)$$
$$\gamma\sqrt{(E'G' - F'^2)} = \sqrt{EE'}\,.\sin(\psi - \omega)$$
$$\delta\sqrt{(E'G' - F'^2)} = \sqrt{GE'}\,.\sin\psi$$

Since by the substitutions

$$dp' = \alpha\,dp + \beta\,dq,$$
$$dq' = \gamma\,dp + \delta\,dq$$

the trinomial

$$E' dp'^2 + 2 F' dp' . dq' + G' dq'^2$$

is transformed into

$$E dp^2 + 2 F dp . dq + G dq^2,$$

we easily obtain

$$E G - F^2 = (E' G' - F'^2)(\alpha\delta - \beta\gamma)^2$$

and since, *vice versa*, the latter trinomial must be transformed into the former by the substitution

$$(\alpha\delta - \beta\gamma) dp = \delta dp' - \beta dq', \quad (\alpha\delta - \beta\gamma) dq = -\gamma dp' + \alpha dq',$$

we find

$$E\delta^2 - 2F\gamma\delta + G\gamma^2 = \frac{EG - F^2}{E'G' - F'^2} \cdot E'$$

$$-E\beta\delta + F(\alpha\delta + \beta\gamma) - G\alpha\gamma = \frac{EG - F^2}{E'G' - F'^2} \cdot F' \qquad 21$$

$$E\beta^2 - 2F\alpha\beta + G\alpha^2 = \frac{EG - F^2}{E'G' - F'^2} \cdot G'$$

22.

From the general discussion of the preceding article we proceed to the very extended application in which, while keeping for p, q their most general meaning, we take for p', q' the quantities denoted in Art. 15 by r, ϕ. We shall use r, ϕ here also in such a way that, for any point whatever on the surface, r will be the shortest distance from a fixed point, and ϕ the angle at this point between the first element of r and a fixed direction. We have thus

$$E' = 1, \quad F' = 0, \quad \omega' = 90°.$$

Let us set also

$$\sqrt{G'} = m,$$

so that any linear element whatever becomes equal to

$$\sqrt{(dr^2 + m^2 d\phi^2)}.$$

Consequently, the four equations deduced in the preceding article for α, β, γ, δ give

$$\sqrt{E} . \cos(\omega - \psi) = \frac{\partial r}{\partial p} \quad . \quad . \quad . \quad . \quad . \quad . \quad . \quad . \quad (1)$$

$$\sqrt{G} . \cos\psi = \frac{\partial r}{\partial q} \quad . \quad . \quad . \quad . \quad . \quad . \quad . \quad . \quad . \quad (2)$$

$$\sqrt{E} . \sin(\psi - \omega) = m . \frac{\partial \phi}{\partial p} \quad . \quad . \quad . \quad . \quad . \quad . \quad . \quad (3)$$

$$\sqrt{G} . \sin \psi = m . \frac{\partial \phi}{\partial q} \quad . \quad . \quad . \quad . \quad . \quad . \quad . \quad . \quad . \quad (4)$$

But the last and the next to the last equations of the preceding article give

$$EG - F^2 = E\left(\frac{\partial r}{\partial q}\right)^2 - 2F \cdot \frac{\partial r}{\partial p} \cdot \frac{\partial r}{\partial q} + G\left(\frac{\partial r}{\partial p}\right)^2 \quad . \quad . \quad (5)$$

$$\left(E \cdot \frac{\partial r}{\partial q} - F \cdot \frac{\partial r}{\partial q}\right) \cdot \frac{\partial \phi}{\partial q} = \left(F \cdot \frac{\partial r}{\partial q} - G \cdot \frac{\partial r}{\partial p}\right) \cdot \frac{\partial \phi}{\partial p} \quad . \quad (6)$$

From these equations must be determined the quantities r, ϕ, ψ and (if need be) m, as functions of p and q. Indeed, integration of equation (5) will give r; r being found, integration of equation (6) will give ϕ; and one or other of equations (1), (2) will give ψ itself. Finally, m is obtained from one or other of equations (3), (4).

The general integration of equations (5), (6) must necessarily introduce two arbitrary functions. We shall easily understand what their meaning is, if we remember that these equations are not limited to the case we are here considering, but are equally valid if r and ϕ are taken in the more general sense of Art. 16, so that r is the length of the shortest line drawn normal to a fixed but arbitrary line, and ϕ is an arbitrary function of the length of that part of the fixed line which is intercepted between any shortest line and an arbitrary fixed point. The general solution must embrace all this in a general way, and the arbitrary functions must go over into definite functions only when the arbitrary line and the arbitrary functions of its parts, which ϕ must represent, are themselves defined. In our case an infinitely small circle may be taken, having its centre at the point from which the distances r are measured, and ϕ will denote the parts themselves of this circle, divided by the radius. Whence it is easily seen that the equations (5), (6) are quite sufficient for our case, provided that the functions which they leave undefined satisfy the condition which r and ϕ satisfy for the initial point and for points at an infinitely small distance from this point.

Moreover, in regard to the integration itself of the equations (5), (6), we know that it can be reduced to the integration of ordinary differential equations, which, however, often happen to be so complicated that there is little to be gained by the reduction. On the contrary, the development in series, which are abundantly sufficient for practical requirements, when only a finite portion of the surface is under consideration, presents no difficulty; and the formulæ thus derived open a fruitful source for the solution of many important problems. But here we shall develop only a single example in order to show the nature of the method.

23.[22]

We shall now consider the case where all the lines for which p is constant are shortest lines cutting orthogonally the line for which $\phi = 0$, which line we can regard as the axis of abscissas. Let A be the point for which $r = 0$, D any point whatever on the axis of abscissas, $AD = p$, B any point whatever on the shortest line normal to AD at D, and $BD = q$, so that p can be regarded as the abscissa, q the ordinate of the point B. The abscissas we assume positive on the branch of the axis of abscissas to which $\phi = 0$ corresponds, while we always regard r as positive. We take the ordinates positive in the region in which ϕ is measured between 0 and 180°.

By the theorem of Art. 16 we shall have

$$\omega = 90^\circ, \quad F = 0, \quad G = 1,$$

and we shall set also

$$\sqrt{E} = n.$$

Thus n will be a function of p, q, such that for $q = 0$ it must become equal to unity. The application of the formula of Art. 18 to our case shows that on any shortest line *whatever* we must have

$$d\theta = \frac{\partial n}{\partial q} \cdot dp,$$ [23]

where θ denotes the angle between the element of this line and the element of the line for which q is constant. Now since the axis of abscissas is itself a shortest line, and since, for it, we have everywhere $\theta = 0$, we see that for $q = 0$ we must have everywhere

$$\frac{\partial n}{\partial q} = 0.$$

Therefore we conclude that, if n is developed into a series in ascending powers of q, this series must have the following form:

$$n = 1 + fq^2 + gq^3 + hq^4 + \text{etc.}$$

where f, g, h, etc., will be functions of p, and we set

$$f = f^\circ + f'p + f''p^2 + \text{etc.}$$
$$g = g^\circ + g'p + g''p^2 + \text{etc.}$$
$$h = h^\circ + h'p + h''p^2 + \text{etc.}$$

or

$$\begin{aligned} n = 1 + f^\circ q^2 + f' p q^2 + f'' p^2 q^2 &+ \text{etc.} \\ + g^\circ q^3 + g' p q^3 &+ \text{etc.} \\ + h^\circ q^4 &+ \text{etc. etc.} \end{aligned}$$

24.

The equations of Art. 22 give, in our case,

$$n \sin\psi = \frac{\partial r}{\partial p}, \quad \cos\psi = \frac{\partial r}{\partial q}, \quad -n\cos\psi = m \,.\, \frac{\partial \phi}{\partial q}, \quad \sin\psi = m \,.\, \frac{\partial \phi}{\partial q},$$

$$n^2 = n^2 \left(\frac{\partial r}{\partial q}\right)^2 + \left(\frac{\partial r}{\partial p}\right)^2, \quad n^2 . \frac{\partial r}{\partial q} \cdot \frac{\partial \phi}{\partial q} + \frac{\partial r}{\partial p} \cdot \frac{\partial \phi}{\partial p} = 0$$

By the aid of these equations, the fifth and sixth of which are contained in the others, series can be developed for r, ϕ, ψ, m, or for any functions whatever of these quantities. We are going to establish here those series that are especially worthy of attention.

Since for infinitely small values of p, q we must have

$$r^2 = p^2 + q^2,$$

the series for r^2 will begin with the terms $p^2 + q^2$. We obtain the terms of higher order by the method of undetermined coefficients,* by means of the equation

$$\left(\frac{1}{n} \cdot \frac{\partial (r^2)}{\partial p}\right)^2 + \left(\frac{\partial (r^2)}{\partial q}\right)^2 = 4\, r^2$$

Thus we have

[1] [24]
$$\begin{aligned} r^2 = p^2 &+ \tfrac{2}{3} f^\circ p^2 q^2 + \tfrac{1}{2} f' p^3 q^2 + (\tfrac{2}{5} f'' - \tfrac{4}{45} f^{\circ 2}) p^4 q^2 \quad \text{etc.} \\ + q^2 & \qquad\qquad + \tfrac{1}{2} g^\circ p^2 q^3 + \tfrac{2}{5} g' p^3 q^3 \\ & \qquad\qquad\qquad + (\tfrac{2}{5} h^\circ - \tfrac{7}{45} f^{\circ 2}) p^2 q^4 \end{aligned}$$

Then we have, from the formula

$$r \sin\psi = \frac{1}{2n} \cdot \frac{\partial (r^2)}{\partial p},$$

[2] [25]
$$\begin{aligned} r \sin\psi = p - \tfrac{1}{3} f^\circ p q^2 - \tfrac{1}{4} f' p^2 q^2 &- (\tfrac{1}{5} f'' + \tfrac{8}{45} f^{\circ 2}) p^3 q^2 \quad \text{etc.} \\ - \tfrac{1}{2} g^\circ p q^3 &- \tfrac{2}{5} g' p^2 q^3 \\ &- (\tfrac{3}{5} h^\circ - \tfrac{8}{45} f^{\circ 2}) p q^4 \end{aligned}$$

---

* We have thought it useless to give the calculation here, which can be somewhat abridged by certain artifices.

and from the formula

$$r\cos\psi = \tfrac{1}{2}\frac{\partial(r^2)}{\partial q}$$

[3] [26]

$$\begin{aligned} r\cos\psi = q &+ \tfrac{2}{3}f^\circ p^2 q + \tfrac{1}{2}f'\, p^3 q + (\tfrac{2}{5}f'' - \tfrac{4}{45}f^{\circ 2})p^4 q \quad \text{etc.} \\ &+ \tfrac{3}{4}g^\circ p^2 q^2 + \tfrac{3}{5}g'\, p^3 q^2 \\ &\qquad + (\tfrac{4}{5}h^\circ - \tfrac{14}{45}f^{\circ 2})p^2 q^3 \end{aligned}$$

These formulæ give the angle ψ. In like manner, for the calculation of the angle ϕ, series for $r\cos\phi$ and $r\sin\phi$ are very elegantly developed by means of the partial differential equations

$$\frac{\partial\,.\,r\cos\phi}{\partial p} = n\cos\phi\,.\,\sin\psi - r\sin\phi\,.\,\frac{\partial\phi}{\partial p}$$

$$\frac{\partial\,.\,r\cos\phi}{\partial q} = \cos\phi\,.\,\cos\psi - r\sin\phi\,.\,\frac{\partial\phi}{\partial q}$$

$$\frac{\partial\,.\,r\sin\phi}{\partial p} = n\sin\phi\,.\,\sin\psi + r\cos\phi\,.\,\frac{\partial\phi}{\partial p}$$

$$\frac{\partial\,.\,r\sin\phi}{\partial q} = \sin\phi\,.\,\cos\psi + r\cos\phi\,.\,\frac{\partial\phi}{\partial q}$$

$$n\cos\psi\,.\,\frac{\partial\phi}{\partial q} + \sin\psi\,.\,\frac{\partial\phi}{\partial p} = 0$$

A combination of these equations gives

$$\frac{r\sin\psi}{n}\,.\,\frac{\partial\,.\,r\cos\phi}{\partial p} + r\cos\psi\,.\,\frac{\partial\,.\,r\cos\phi}{\partial q} = r\cos\phi$$

$$\frac{r\sin\psi}{n}\,.\,\frac{\partial\,.\,r\sin\phi}{\partial p} + r\cos\psi\,.\,\frac{\partial\,.\,r\sin\phi}{\partial q} = r\sin\phi$$

From these two equations series for $r\cos\phi$, $r\sin\phi$ are easily developed, whose first terms must evidently be p, q respectively. The series are

[4] [27]

$$\begin{aligned} r\cos\phi = p &+ \tfrac{2}{3}f^\circ p q^2 + \tfrac{5}{12}f'\, p^2 q^2 + (\tfrac{3}{10}f'' - \tfrac{8}{45}f^{\circ 2})p^3 q^2 \quad \text{etc.} \\ &+ \tfrac{1}{2}g^\circ p q^3 + \tfrac{7}{20}g'\, p^2 q^3 \\ &\qquad + (\tfrac{2}{5}h^\circ - \tfrac{7}{45}f^{\circ 2})p q^4 \end{aligned}$$

[5] [28]

$$\begin{aligned} r\sin\phi = q &- \tfrac{1}{3}f^\circ p^2 q - \tfrac{1}{6}f'\, p^3 q - (\tfrac{1}{10}f'' - \tfrac{7}{90}f^{\circ 2})p^4 q \quad \text{etc.} \\ &- \tfrac{1}{4}g^\circ p^2 q^2 - \tfrac{3}{20}g'\, p^3 q^2 \\ &\qquad - (\tfrac{1}{5}h^\circ + \tfrac{13}{90}f^{\circ 2})p^2 q^3 \end{aligned}$$

From a combination of equations [2], [3], [4], [5] a series for $r^2\cos(\psi+\phi)$, may be derived, and from this, dividing by the series [1], a series for $\cos(\psi+\phi)$, from which may be found a series for the angle $\psi+\phi$ itself. However, the same series can be obtained more elegantly in the following manner. By differentiating the first and second of the equations introduced at the beginning of this article, we obtain

$$\sin\psi\cdot\frac{\partial n}{\partial q}+n\cos\psi\cdot\frac{\partial\psi}{\partial q}+\sin\psi\cdot\frac{\partial\psi}{\partial q}=0$$

and this combined with the equation

$$n\cos\psi\cdot\frac{\partial\phi}{\partial q}+\sin\psi\cdot\frac{\partial\phi}{\partial p}=0$$

gives

$$\frac{r\sin\psi}{n}\cdot\frac{\partial n}{\partial q}+\frac{r\sin\psi}{n}\cdot\frac{\partial(\psi+\phi)}{\partial p}+r\cos\psi\cdot\frac{\partial(\psi+\phi)}{\partial q}=0$$

From this equation, by aid of the method of undetermined coefficients, we can easily derive the series for $\psi+\phi$, if we observe that its first term must be $\frac{1}{2}\pi$, the radius being taken equal to unity and 2π denoting the circumference of the circle,

$$[6]\qquad \begin{aligned}\psi+\phi=\tfrac{1}{2}\pi &-f^\circ pq-\tfrac{2}{3}f'p^2q-(\tfrac{1}{2}f''-\tfrac{1}{6}f^{\circ 2})p^3q \quad \text{etc.}\\ &-g^\circ pq^2-\tfrac{3}{4}g'p^2q^2\\ &-(h^\circ-\tfrac{1}{3}f^{\circ 2})pq^3\end{aligned}$$

It seems worth while also to develop the area of the triangle ABD into a series. For this development we may use the following conditional equation, which is easily derived from sufficiently obvious geometric considerations, and in which S denotes the required area:

$$\frac{r\sin\psi}{n}\cdot\frac{\partial S}{\partial p}+r\cos\psi\cdot\frac{\partial S}{\partial q}=\frac{r\sin\psi}{n}\cdot\int n\,dq$$ [29]

the integration beginning with $q=0$. From this equation we obtain, by the method of undetermined coefficients,

$$[7]^{30}\qquad \begin{aligned}S=\tfrac{1}{2}pq &-\tfrac{1}{12}f^\circ p^3q-\tfrac{1}{20}f'p^4q-(\tfrac{1}{30}f''-\tfrac{1}{60}f^{\circ 2})p^5q \quad \text{etc.}\\ &-\tfrac{1}{12}f^\circ pq^3-\tfrac{3}{40}g^\circ p^3q^2-\tfrac{1}{20}g'p^4q^2\\ &-\tfrac{7}{120}f'p^2q^3-(\tfrac{1}{15}h^\circ+\tfrac{2}{45}f''+\tfrac{1}{60}f^{\circ 2})p^3q^3\\ &-\tfrac{1}{10}g^\circ pq^4-\tfrac{3}{40}g'p^2q^4\\ &-(\tfrac{1}{10}h^\circ-\tfrac{1}{30}f^{\circ 2})pq^5\end{aligned}$$

25.

From the formulæ of the preceding article, which refer to a right triangle formed by shortest lines, we proceed to the general case. Let C be another point on the same shortest line DB, for which point p remains the same as for the point B, and q', r', ϕ', ψ', S' have the same meanings as q, r, ϕ, ψ, S have for the point B. There will thus be a triangle between the points A, B, C, whose angles we denote by A, B, C, the sides opposite these angles by a, b, c, and the area by σ. We represent the measure of curvature at the points A, B, C by α, β, γ respectively. And then supposing (which is permissible) that the quantities $p, q, q-q'$ are positive, we shall have

$$A=\phi-\phi', \quad B=\psi, \quad C=\pi-\psi',$$
$$a=q-q', \quad b=r', \quad c=r, \qquad \sigma=S-S'.$$

We shall first express the area σ by a series. By changing in [7] each of the quantities that refer to B into those that refer to C, we obtain a formula for S'. Whence we have, exact to quantities of the sixth order,

$$\begin{aligned}\sigma=\tfrac{1}{2}p(q-q')\big(1-\tfrac{1}{6}f^\circ(p^2+q^2+qq'+q'^2)\\ -\tfrac{1}{60}f'p(6p^2+7q^2+7qq'+7q'^2)\\ -\tfrac{1}{20}g^\circ(q+q')(3p^2+4q^2+4q'^2)\big)\end{aligned} \qquad 31$$

This formula, by aid of series [2], namely,

$$c\sin B=p(1-\tfrac{1}{3}f^\circ q^2-\tfrac{1}{4}f'pq^2-\tfrac{1}{2}g^\circ q^3-\text{etc.})$$

can be changed into the following:

$$\begin{aligned}\sigma=\tfrac{1}{2}ac\sin B\big(1-\tfrac{1}{6}f^\circ(p^2-q^2+qq'+q'^2)\\ -\tfrac{1}{60}f'p(6p^2-8q^2+7qq'+7q'^2)\\ -\tfrac{1}{20}g^\circ(3p^2q+3p^2q'-6p^3+4q^2q'+4qq'^2+4q'^3)\big)\end{aligned}$$

The measure of curvature for any point whatever of the surface becomes (by Art. 19, where m, p, q were what n, q, p are here)

$$\begin{aligned}k=-\frac{1}{n}\cdot\frac{\partial^2 n}{\partial q^2}=-\frac{2f+6gq+12hq^2+\text{etc.}}{1+fq^2+\text{etc.}}\\ =-2f-6gq-(12h-2f^2)q^2-\text{etc.}\end{aligned}$$

Therefore we have, when p, q refer to the point B,

$$\beta=-2f^\circ-2f'p-6g^\circ q-2f''p^2-6g'pq-(12h^\circ-2f^{\circ 2})q^2-\text{etc.}$$

Also

$$\gamma = -2f^\circ - 2f'p - 6g^\circ q' - 2f''p^2 - 6g'pq' - (12h^\circ - 2f^{\circ 2})q'^2 - \text{etc.}$$
$$\alpha = -2f^\circ$$

Introducing these measures of curvature into the expression for σ, we obtain the following expression, exact to quantities of the sixth order (exclusive):

$$\begin{aligned}\sigma = \tfrac{1}{2}ac\sin B\,\big(1 &+ \tfrac{1}{120}\alpha(4p^2 - 2q^2 + 3qq' + 3q'^2)\\ &+ \tfrac{1}{120}\beta(3p^2 - 6q^2 + 6qq' + 3q'^2)\\ &+ \tfrac{1}{120}\gamma(3p^2 - 2q^2 + qq' + 4q'^2)\big)\end{aligned}$$ 32

The same precision will remain, if for p, q, q' we substitute $c\sin B$, $c\cos B$, $c\cos B - a$. This gives

[8] 33

$$\begin{aligned}\sigma = \tfrac{1}{2}ac\sin B\,\big(1 &+ \tfrac{1}{120}\alpha(3a^2 + 4c^2 - 9ac\cos B)\\ &+ \tfrac{1}{120}\beta(3a^2 + 3c^2 - 12ac\cos B)\\ &+ \tfrac{1}{120}\gamma(4a^2 + 3c^2 - 9ac\cos B)\big)\end{aligned}$$

Since all expressions which refer to the line AD drawn normal to BC have disappeared from this equation, we may permute among themselves the points A, B, C and the expressions that refer to them. Therefore we shall have, with the same precision,

[9]

$$\begin{aligned}\sigma = \tfrac{1}{2}bc\sin A\,\big(1 &+ \tfrac{1}{120}\alpha(3b^2 + 3c^2 - 12bc\cos A)\\ &+ \tfrac{1}{120}\beta(3b^2 + 4c^2 - 9bc\cos A)\\ &+ \tfrac{1}{120}\gamma(4b^2 + 3c^2 - 9bc\cos A)\big)\end{aligned}$$

[10]

$$\begin{aligned}\sigma = \tfrac{1}{2}ab\sin C\,\big(1 &+ \tfrac{1}{120}\alpha(3a^2 + 4b^2 - 9ab\cos C)\\ &+ \tfrac{1}{120}\beta(4a^2 + 3b^2 - 9ab\cos C)\\ &+ \tfrac{1}{120}\gamma(3a^2 + 3b^2 - 12ab\cos C)\big)\end{aligned}$$

26.

The consideration of the rectilinear triangle whose sides are equal to a, b, c is of great advantage. The angles of this triangle, which we shall denote by A^*, B^*, C^*, differ from the angles of the triangle on the curved surface, namely, from A, B, C, by quantities of the second order; and it will be worth while to develop these differences accurately. However, it will be sufficient to show the first steps in these more tedious than difficult calculations.

Replacing in formulæ [1], [4], [5] the quantities that refer to B by those that refer to C, we get formulæ for r'^2, $r'\cos\phi'$, $r'\sin\phi'$. Then the development of the expression

$$r^2+r'^2-(q-q')^2-2r\cos\phi\,.\,r'\cos\phi'-2r\sin\phi\,.\,r'\sin\phi'$$
$$=b^2+c^2-a^2-2bc\cos A$$
$$=2bc(\cos A^*-\cos A),$$

combined with the development of the expression

$$r\sin\phi\,.\,r'\cos\phi'-r\cos\phi\,.\,r'\sin\phi'=bc\sin A,$$

gives the following formula:

$$\cos A^*-\cos A=-(q-q')p\sin A\left(\tfrac{1}{3}f^\circ+\tfrac{1}{6}f'p+\tfrac{1}{4}g^\circ(q+q')\right.$$
$$+(\tfrac{1}{10}f''-\tfrac{1}{45}f^{\circ 2})p^2+\tfrac{3}{20}g'p(q+q')$$
$$\left.+(\tfrac{1}{5}h^\circ-\tfrac{7}{90}f^{\circ 2})(q^2+qq'+q'^2)+\text{etc.}\right)$$

From this we have, to quantities of the fifth order,

$$A^*-A=+(q-q')p\left(\tfrac{1}{3}f^\circ+\tfrac{1}{6}f'p+\tfrac{1}{4}g^\circ(q+q')+\tfrac{1}{10}f''p^2\right.$$
$$+\tfrac{3}{20}g'p(q+q')+\tfrac{1}{5}h^\circ(q^2+qq'+q'^2)$$
$$\left.-\tfrac{1}{90}f^{\circ 2}(7p^2+7q^2+12qq'+7q'^2)\right) \qquad 34$$

Combining this formula with

$$2\sigma=ap\left(1-\tfrac{1}{6}f^\circ(p^2+q^2+qq'+q'^2)-\text{etc.}\right)$$

and with the values of the quantities α, β, γ found in the preceding article, we obtain, to quantities of the fifth order,

[11] [35]
$$A^*=A-\sigma\left(\tfrac{1}{6}\alpha+\tfrac{1}{12}\beta+\tfrac{1}{12}\gamma+\tfrac{2}{15}f''p^2+\tfrac{1}{5}g'p(q+q')\right.$$
$$+\tfrac{1}{5}h^\circ(3q^2-2qq'+3q'^2)$$
$$\left.+\tfrac{1}{90}f^{\circ 2}(4p^2-11q^2+14qq'-11q'^2)\right)$$

By precisely similar operations we derive

[12] [36]
$$B^*=B-\sigma\left(\tfrac{1}{12}\alpha+\tfrac{1}{6}\beta+\tfrac{1}{12}\gamma+\tfrac{1}{10}f''p^2+\tfrac{1}{10}g'p(2q+q')\right.$$
$$+\tfrac{1}{5}h^\circ(4q^2-4qq'+3q'^2)$$
$$\left.-\tfrac{1}{90}f^{\circ 2}(2p^2+8q^2-8qq'+11q'^2)\right)$$

[13] [37]
$$C^*=C-\sigma\left(\tfrac{1}{12}\alpha+\tfrac{1}{12}\beta+\tfrac{1}{6}\gamma+\tfrac{1}{10}f''p^2+\tfrac{1}{10}g'p(q+2q')\right.$$
$$+\tfrac{1}{5}h^\circ(3q^2-4qq'+4q'^2)$$
$$\left.-\tfrac{1}{90}f^{\circ 2}(2p^2+11q^2-8qq'+8q'^2)\right)$$

From these formulæ we deduce, since the sum $A^*+B^*+C^*$ is equal to two right angles, the excess of the sum $A+B+C$ over two right angles, namely,

[14] [38]
$$A+B+C=\pi+\sigma\left(\tfrac{1}{3}\alpha+\tfrac{1}{3}\beta+\tfrac{1}{3}\gamma+\tfrac{1}{3}f''p^2+\tfrac{1}{2}g'p(q+q')\right.$$
$$\left.+(2h^\circ-\tfrac{1}{3}f^{\circ 2})(q^2-qq'+q'^2)\right)$$

This last equation could also have been derived from formula [6].

27.

If the curved surface is a sphere of radius R, we shall have

$$\alpha=\beta=\gamma=-2f^\circ=\frac{1}{R^2};\ f''=0,\quad g'=0,\quad 6h^\circ-f^{\circ 2}=0,$$

or

$$h^\circ=\frac{1}{24R^4}.$$

Consequently, formula [14] becomes

$$A+B+C=\pi+\frac{\sigma}{R^2},$$

which is absolutely exact. But formulæ [11], [12], [13] give

$$A^*=A-\frac{\sigma}{3R^2}-\frac{\sigma}{180R^4}(2p^2-q^2+4qq'-q'^2)$$

$$B^*=B-\frac{\sigma}{3R^2}+\frac{\sigma}{180R^4}(p^2-2q^2+2qq'+q'^2)$$

$$C^*=C-\frac{\sigma}{3R^2}+\frac{\sigma}{180R^4}(p^2+q^2+2qq'-2q'^2)$$

or, with equal exactness,

$$A^*=A-\frac{\sigma}{3R^2}-\frac{\sigma}{180R^4}(b^2+c^2-2a^2)$$

$$B^*=B-\frac{\sigma}{3R^2}-\frac{\sigma}{180R^4}(a^2+c^2-2b^2)$$

$$C^*=C-\frac{\sigma}{3R^2}-\frac{\sigma}{180R^4}(a^2+b^2-2c^2)$$

Neglecting quantities of the fourth order, we obtain from the above the well-known theorem first established by the illustrious Legendre. [39]

28.

Our general formulæ, if we neglect terms of the fourth order, become extremely simple, namely:

$$A^*=A-\tfrac{1}{12}\sigma(2\alpha+\beta+\gamma)$$
$$B^*=B-\tfrac{1}{12}\sigma(\alpha+2\beta+\gamma)$$
$$C^*=C-\tfrac{1}{12}\sigma(\alpha+\beta+2\gamma)$$

Thus to the angles A, B, C on a non-spherical surface, unequal reductions must be applied, so that the sines of the changed angles become proportional to the sides opposite. The inequality, generally speaking, will be of the third order; but if the surface differs little from a sphere, the inequality will be of a higher order. Even in the greatest triangles on the earth's surface, whose angles it is possible to measure, the difference can always be regarded as insensible. Thus, *e. g.*, in the greatest of the triangles which we have measured in recent years, namely, that between the points Hohehagen, Brocken, Inselberg,[40] where the excess of the sum of the angles was 14″.85348, the calculation gave the following reductions to be applied to the angles:

Hohehagen	−4″.95113
Brocken	−4″.95104
Inselberg	−4″.95131.

29.[41]

We shall conclude this study by comparing the area of a triangle on a curved surface with the area of the rectilinear triangle whose sides are a, b, c. We shall denote the area of the latter by σ^*; hence

$$\sigma^* = \tfrac{1}{2}\, bc \sin A^* = \tfrac{1}{2}\, ac \sin B^* = \tfrac{1}{2}\, ab \sin C^*$$

We have, to quantities of the fourth order,

$$\sin A^* = \sin A - \tfrac{1}{12}\,\sigma \cos A\,.\,(2\,\alpha + \beta + \gamma)$$

or, with equal exactness,

$$\sin A = \sin A^*\,.\,(1 + \tfrac{1}{24}\, bc \cos A\,.\,(2\,\alpha + \beta + \gamma))$$

Substituting this value in formula [9], we shall have, to quantities of the sixth order,

$$\begin{aligned}\sigma = \tfrac{1}{2}\, bc \sin A^*\,.\,(1 &+ \tfrac{1}{120}\,\alpha\,(3\,b^2 + 3\,c^2 - 2\,bc \cos A)\\ &+ \tfrac{1}{120}\,\beta\,(3\,b^2 + 4\,c^2 - 4\,bc \cos A)\\ &+ \tfrac{1}{120}\,\gamma\,(4\,b^2 + 3\,c^2 - 4\,bc \cos A)),\end{aligned}$$

or, with equal exactness,

$$\sigma = \sigma^*\,(1 + \tfrac{1}{120}\,\alpha\,(a^2 + 2\,b^2 + 2\,c^2) + \tfrac{1}{120}\,\beta\,(2\,a^2 + b^2 + 2\,c^2) + \tfrac{1}{120}\,\gamma\,(2\,a^2 + 2\,b^2 + c^2))$$

For the sphere this formula goes over into the following form:

$$\sigma = \sigma^*\,(1 + \tfrac{1}{24}\,\alpha\,(a^2 + b^2 + c^2)).$$

It is easily verified that, with the same precision, the following formula may be taken instead of the above:

$$\sigma = \sigma^* \sqrt{\frac{\sin A \ . \sin B \ . \sin C}{\sin A^* . \sin B^* . \sin C^*}}$$

If this formula is applied to triangles on non-spherical curved surfaces, the error, generally speaking, will be of the fifth order, but will be insensible in all triangles such as may be measured on the earth's surface.

GAUSS'S ABSTRACT OF THE DISQUISITIONES GENERALES CIRCA SUPERFICIES CURVAS, PRESENTED TO THE ROYAL SOCIETY OF GÖTTINGEN.

GÖTTINGISCHE GELEHRTE ANZEIGEN. No. 177. PAGES 1761–1768. 1827. NOVEMBER 5.

On the 8th of October, Hofrath Gauss presented to the Royal Society a paper: *Disquisitiones generales circa superficies curvas.*

Although geometers have given much attention to general investigations of curved surfaces and their results cover a significant portion of the domain of higher geometry, this subject is still so far from being exhausted, that it can well be said that, up to this time, but a small portion of an exceedingly fruitful field has been cultivated. Through the solution of the problem, to find all representations of a given surface upon another in which the smallest elements remain unchanged, the author sought some years ago to give a new phase to this study. The purpose of the present discussion is further to open up other new points of view and to develop some of the new truths which thus become accessible. We shall here give an account of those things which can be made intelligible in a few words. But we wish to remark at the outset that the new theorems as well as the presentations of new ideas, if the greatest generality is to be attained, are still partly in need of some limitations or closer determinations, which must be omitted here.

In researches in which an infinity of directions of straight lines in space is concerned, it is advantageous to represent these directions by means of those points upon a fixed sphere, which are the end points of the radii drawn parallel to the lines. The centre and the radius of this *auxiliary sphere* are here quite arbitrary. The radius may be taken equal to unity. This procedure agrees fundamentally with that which is constantly employed in astronomy, where all directions are referred to a fictitious celestial sphere of infinite radius. Spherical trigonometry and certain other theorems, to which the author has added a new one of frequent application, then serve for the solution of the problems which the comparison of the various directions involved can present.

If we represent the direction of the normal at each point of the curved surface by the corresponding point of the sphere, determined as above indicated, namely, in this way, to every point on the surface, let a point on the sphere correspond; then, generally speaking, to every line on the curved surface will correspond a line on the sphere, and to every part of the former surface will correspond a part of the latter. The less this part differs from a plane, the smaller will be the corresponding part on the sphere. It is, therefore, a very natural idea to use as the measure of the total curvature, which is to be assigned to a part of the curved surface, the area of the corresponding part of the sphere. For this reason the author calls this area the *integral curvature* of the corresponding part of the curved surface. Besides the magnitude of the part, there is also at the same time its *position* to be considered. And this position may be in the two parts similar or inverse, quite independently of the relation of their magnitudes. The two cases can be distinguished by the positive or negative sign of the total curvature. This distinction has, however, a definite meaning only when the figures are regarded as upon definite sides of the two surfaces. The author regards the figure in the case of the sphere on the outside, and in the case of the curved surface on that side upon which we consider the normals erected. It follows then that the positive sign is taken in the case of convexo-convex or concavo-concave surfaces (which are not essentially different), and the negative in the case of concavo-convex surfaces. If the part of the curved surface in question consists of parts of these different sorts, still closer definition is necessary, which must be omitted here.

The comparison of the areas of two corresponding parts of the curved surface and of the sphere leads now (in the same manner as, *e. g.*, from the comparison of volume and mass springs the idea of density) to a new idea. The author designates as *measure of curvature* at a point of the curved surface the value of the fraction whose denominator is the area of the infinitely small part of the curved surface at this point and whose numerator is the area of the corresponding part of the surface of the auxiliary sphere, or the integral curvature of that element. It is clear that, according to the idea of the author, integral curvature and measure of curvature in the case of curved surfaces are analogous to what, in the case of curved lines, are called respectively amplitude and curvature simply. He hesitates to apply to curved surfaces the latter expressions, which have been accepted more from custom than on account of fitness. Moreover, less depends upon the choice of words than upon this, that their introduction shall be justified by pregnant theorems.

The solution of the problem, to find the measure of curvature at any point of a curved surface, appears in different forms according to the manner in which the nature of the curved surface is given. When the points in space, in general, are distinguished by three rectangular coordinates, the simplest method is to express one coordinate as a function of the other two. In this way we obtain the simplest expression for the measure of curvature. But, at the same time, there arises a remarkable relation between this measure of curvature and the curvatures of the curves formed by the intersections of the curved surface with planes normal to it. EULER, as is well known, first showed that two of these cutting planes which intersect each other at right angles have this property, that in one is found the greatest and in the other the smallest radius of curvature; or, more correctly, that in them the two extreme curvatures are found. It will follow then from the above mentioned expression for the measure of curvature that this will be equal to a fraction whose numerator is unity and whose denominator is the product of the extreme radii of curvature. The expression for the measure of curvature will be less simple, if the nature of the curved surface is determined by an equation in x, y, z. And it will become still more complex, if the nature of the curved surface is given so that x, y, z are expressed in the form of functions of two new variables p, q. In this last case the expression involves fifteen elements, namely, the partial differential coefficients of the first and second orders of x, y, z with respect to p and q. But it is less important in itself than for the reason that it facilitates the transition to another expression, which must be classed with the most remarkable theorems of this study. If the nature of the curved surface be expressed by this method, the general expression for any linear element upon it, or for $\sqrt{(dx^2+dy^2+dz^2)}$, has the form $\sqrt{(E\,dp^2+2\,F\,dp\,.\,dq+G\,dq^2)}$, where E, F, G are again functions of p and q. The new expression for the measure of curvature mentioned above contains merely these magnitudes and their partial differential coefficients of the first and second order. Therefore we notice that, in order to determine the measure of curvature, it is necessary to know only the general expression for a linear element; the expressions for the coordinates x, y, z are not required. A direct result from this is the remarkable theorem: If a curved surface, or a part of it, can be developed upon another surface, the measure of curvature at every point remains unchanged after the development. In particular, it follows from this further: Upon a curved surface that can be developed upon a plane, the measure of curvature is everywhere equal to zero. From this we derive at once the characteristic equation of surfaces developable upon a plane, namely,

$$\frac{\partial^2 z}{\partial x^2}\cdot\frac{\partial^2 z}{\partial y^2}-\left(\frac{\partial^2 z}{\partial x\,.\,\partial y}\right)^2=0,$$

when z is regarded as a function of x and y. This equation has been known for some time, but according to the author's judgment it has not been established previously with the necessary rigor.

These theorems lead to the consideration of the theory of curved surfaces from a new point of view, where a wide and still wholly uncultivated field is open to investigation. If we consider surfaces not as boundaries of bodies, but as bodies of which one dimension vanishes, and if at the same time we conceive them as flexible but not extensible, we see that two essentially different relations must be distinguished, namely, on the one hand, those that presuppose a definite form of the surface in space; on the other hand, those that are independent of the various forms which the surface may assume. This discussion is concerned with the latter. In accordance with what has been said, the measure of curvature belongs to this case. But it is easily seen that the consideration of figures constructed upon the surface, their angles, their areas and their integral curvatures, the joining of the points by means of shortest lines, and the like, also belong to this case. All such investigations must start from this, that the very nature of the curved surface is given by means of the expression of any linear element in the form $\sqrt{(E\,dp^2 + 2\,F\,dp\,.\,dq + G\,dq^2)}$. The author has embodied in the present treatise a portion of his investigations in this field, made several years ago, while he limits himself to such as are not too remote for an introduction, and may, to some extent, be generally helpful in many further investigations. In our abstract, we must limit ourselves still more, and be content with citing only a few of them as types. The following theorems may serve for this purpose.

If upon a curved surface a system of infinitely many shortest lines of equal lengths be drawn from one initial point, then will the line going through the end points of these shortest lines cut each of them at right angles. If at every point of an arbitrary line on a curved surface shortest lines of equal lengths be drawn at right angles to this line, then will all these shortest lines be perpendicular also to the line which joins their other end points. Both these theorems, of which the latter can be regarded as a generalization of the former, will be demonstrated both analytically and by simple geometrical considerations. *The excess of the sum of the angles of a triangle formed by shortest lines over two right angles is equal to the total curvature of the triangle.* It will be assumed here that that angle ($57°\,17'45''$) to which an arc equal to the radius of the sphere corresponds will be taken as the unit for the angles, and that for the unit of total curvature will be taken a part of the spherical surface, the area of which is a square whose side is equal to the radius of the sphere. Evidently we can express this important theorem thus also: the excess over two right angles of the angles of a triangle formed by shortest lines is to eight right angles as the part of the surface of the auxiliary sphere, which corresponds to it as its integral curvature, is to the whole surface of the sphere. In general, the excess over $2\,n-4$ right angles of the angles of a polygon of n sides, if these are shortest lines, will be equal to the integral curvature of the polygon.

The general investigations developed in this treatise will, in the conclusion, be applied to the theory of triangles of shortest lines, of which we shall introduce only a couple of important theorems. If a, b, c be the sides of such a triangle (they will be regarded as magnitudes of the first order); A, B, C the angles opposite; α, β, γ the measures of curvature at the angular points; σ the area of the triangle, then, to magnitudes of the fourth order, $\frac{1}{3}(\alpha+\beta+\gamma)\sigma$ is the excess of the sum $A+B+C$ over two right angles. Further, with the same degree of exactness, the angles of a plane rectilinear triangle whose sides are a, b, c, are respectively

$$A-\tfrac{1}{12}(2\alpha+\beta+\gamma)\sigma$$
$$B-\tfrac{1}{12}(\alpha+2\beta+\gamma)\sigma$$
$$C-\tfrac{1}{12}(\alpha+\beta+2\gamma)\sigma.$$

We see immediately that this last theorem is a generalization of the familiar theorem first established by LEGENDRE. By means of this theorem we obtain the angles of a plane triangle, correct to magnitudes of the fourth order, if we diminish each angle of the corresponding spherical triangle by one-third of the spherical excess. In the case of non-spherical surfaces, we must apply unequal reductions to the angles, and this inequality, generally speaking, is a magnitude of the third order. However, even if the whole surface differs only a little from the spherical form, it will still involve also a factor denoting the degree of the deviation from the spherical form. It is unquestionably important for the higher geodesy that we be able to calculate the inequalities of those reductions and thereby obtain the thorough conviction that, for all measurable triangles on the surface of the earth, they are to be regarded as quite insensible. So it is, for example, in the case of the greatest triangle of the triangulation carried out by the author. The greatest side of this triangle is almost fifteen geographical* miles, and the excess of the sum of its three angles over two right angles amounts almost to fifteen seconds. The three reductions of the angles of the plane triangle are 4″.95113, 4″.95104, 4″.95131. Besides, the author also developed the missing terms of the fourth order in the above expressions. Those for the sphere possess a very simple form. However, in the case of measurable triangles upon the earth's surface, they are quite insensible. And in the example here introduced they would have diminished the first reduction by only two units in the fifth decimal place and increased the third by the same amount.

---

*This German geographical mile is four minutes of arc at the equator, namely, 7.42 kilometers, and is equal to about 4.6 English statute miles. [Translators.]

NOTES.

1. Art. 1, p. 3, l. 3. Gauss got the idea of using the auxiliary sphere from astronomy. *Cf.* Gauss's Abstract, p. 45.

2. Art. 2, p. 3, l. 2 fr. bot. In the Latin text *situs* is used for the direction or orientation of a plane, the position of a plane, the direction of a line, and the position of a point.

3. Art. 2, p. 4, l. 14. In the Latin texts the notation

$$\cos(1)\,L^2 + \cos(2)\,L^2 + \cos(3)\,L^2 = 1$$

is used. This is replaced in the translations (except Böklen's) by the more recent notation

$$\cos^2(1)\,L + \cos^2(2)\,L + \cos^2(3)\,L = 1.$$

5. Art. 2, p. 4, l. 3 fr. bot. This stands in the original and in Liouville's reprint,

$$\cos A\,(\cos t \sin t' - \sin t \cos t')\,(\cos t'' \sin t''' - \sin t'' \mathit{sin}\, t''').$$

4. Art. 2, pp. 4–6. Theorem VI is original with Gauss, as is also the method of deriving VII. The following figures show the points and lines of Theorems VI and VII:

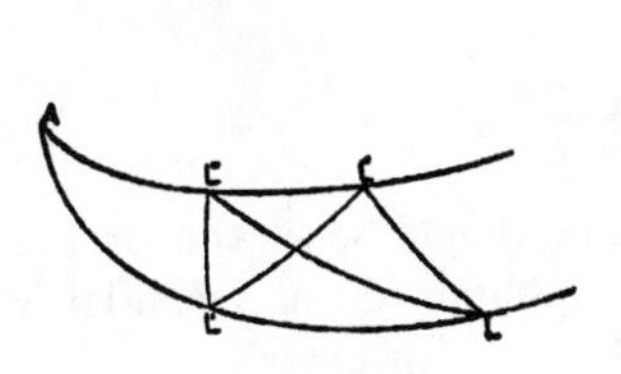

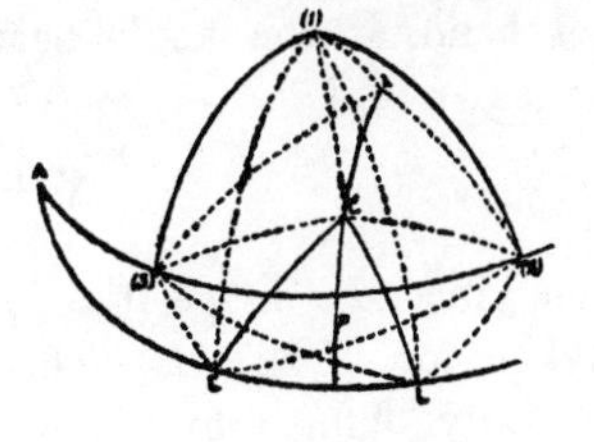

6. Art. 3, p. 6. The geometric condition here stated, that the curvature be continuous for each point of the surface, or part of the surface, considered is equivalent to the analytic condition that the first and second derivatives of the function or functions defining the surface be finite and continuous for all points of the surface, or part of the surface, considered.

7. Art. 4, p. 7, l. 20. In the Latin texts the notation XX for X^2, etc., is used.

8. Art. 4, p. 7. "The second method of representing a surface (the expression of the coordinates by means of two auxiliary variables) was first used by Gauss for arbitrary surfaces in the case of the problem of conformal mapping. [Astronomische Abhandlungen, edited by H. C. Schumacher, vol. III, Altona, 1825; Gauss, Werke, vol. IV, p. 189; reprinted in vol. 55 of Ostwald's Klassiker.—*Cf.* also Gauss, Theoria attractionis corporum sphaer. ellipt., Comment. Gott. II, 1813; Gauss, Werke, vol. V, p. 10.] Here he applies this representation for the first time to the determination of the direction of the surface normal, and later also to the study of curvature and of geodetic lines. The geometrical significance of the variables p, q is discussed more fully in Art. 17. This method of representation forms the source of many new theorems, of which these are particularly worthy of mention: the corollary, that the measure of curvature remains unchanged by the bending of the surface (Art. 11, 12); the theorems of Art. 15, 16 concerning geodetic lines; the theorem of Art. 20; and, finally, the results derived in the conclusion, which refer a geodetic triangle to the rectilinear triangle whose sides are of the same length." [Wangerin.]

9. Art. 5, p. 8. "To decide the question, which of the two systems of values found in Art. 4 for X, Y, Z belong to the normal directed outwards, which to the normal directed inwards, we need only to apply the theorem of Art. 2 (VII), provided we use the second method of representing the surface. If, on the contrary, the surface is defined by the equation between the coordinates $W=0$, then the following simpler con-considerations lead to the answer. We draw the line $d\sigma$ from the point A towards the outer side, then, if dx, dy, dz are the projections of $d\sigma$, we have

$$P\,dx + Q\,dy + R\,dz > 0.$$

On the other hand, if the angle between σ and the normal taken outward is acute, then

$$\frac{dx}{d\sigma}X + \frac{dy}{d\sigma}Y + \frac{dz}{d\sigma}Z > 0.$$

This condition, since $d\sigma$ is positive, must be combined with the preceding, if the first solution is taken for X, Y, Z. This result is obtained in a similar way, if the surface is analytically defined by the third method." [Wangerin.]

10. Art. 6, p. 10, l. 4. The definition of measure of curvature here given is the one generally used. But Sophie Germain defined as a measure of curvature at a point of a surface the sum of the reciprocals of the principal radii of curvature at that point, or double the so-called mean curvature. *Cf.* Crelle's Journ. für Math., vol. VII. Casorati defined as a measure of curvature one-half the sum of the squares of the reciprocals of the principal radii of curvature at a point of the surface. *Cf.* Rend. del R. Istituto Lombardo, ser. 2, vol. 22, 1889; Acta Mathem. vol. XIV, p. 95, 1890.

11. Art. 6, p. 11, l. 21. Gauss did not carry out his intention of studying the most general cases of figures mapped on the sphere.
12. Art. 7, p. 11, l. 31. "That the consideration of a surface element which has the form of a triangle can be used in the calculation of the measure of curvature, follows from this fact that, according to the formula developed on page 12, k is independent of the magnitudes dx, dy, δx, δy, and that, consequently, k has the same value for every infinitely small triangle at the same point of the surface, therefore also for surface elements of any form whatever lying at that point." [Wangerin.]
13. Art. 7, p. 12, l. 20. The notation in the Latin text for the partial derivatives:

$$\frac{dX}{dx}, \frac{dX}{dy}, \text{ etc.},$$

has been replaced throughout by the more recent notation:

$$\frac{\partial X}{\partial x}, \frac{\partial X}{\partial y}, \text{ etc.}$$

14. Art. 7, p. 13, l. 16. This formula, as it stands in the original and in Liouville's reprint, is

$$dY = -Z^3 tu\, dt - Z^3 (1+t^2)\, du.$$

The incorrect sign in the second member has been corrected in the reprint in Gauss, Werke, vol. IV, and in the translations.
15. Art. 8, p. 15, l. 3. Euler's work here referred to is found in Mem. de l'Acad. de Berlin, vol. XVI, 1760.
16. Art. 10, p. 18, ll. 8, 9, 10. Instead of D, D', D'' as here defined, the Italian geometers have introduced magnitudes denoted by the same letters and equal, in Gauss's notation, to

$$\frac{D}{\sqrt{(EG-F^2)}}, \quad \frac{D'}{\sqrt{(EG-F^2)}}, \quad \frac{D''}{\sqrt{(EG-F^2)}}$$

respectively.
17. Art. 11, p. 19, ll. 4, 6, fr. bot. In the original and in Liouville's reprint, two of these formulæ are incorrectly given:

$$\frac{\partial F}{\partial q} = m'' + n, \quad n = \frac{\partial F}{\partial q} - \frac{1}{2}\cdot\frac{\partial E}{\partial q}.$$

The proper corrections have been made in Gauss, Werke, vol. IV, and in the translations.
18. Art. 13, p. 21, l. 20. Gauss published nothing further on the properties of developable surfaces.

19. Art. 14, p. 22, l. 8. The transformation is easily made by means of integration by parts.

20. Art. 17, p. 25. If we go from the point p, q to the point $(p+dp, q)$, and if the Cartesian coordinates of the first point are x, y, z, and of the second $x+dx, y+dy, z+dz$; with ds the linear element between the two points, then the direction cosines of ds are

$$\cos\alpha=\frac{dx}{ds},\quad \cos\beta=\frac{dy}{ds},\quad \cos\gamma=\frac{dz}{ds}.$$

Since we assume here $q=$Constant or $dq=0$, we have also

$$dx=\frac{\partial x}{\partial p}\cdot dp,\quad dy=\frac{\partial y}{\partial p}\cdot dp,\quad dz=\frac{\partial z}{\partial p}\cdot dp,\quad ds=\pm\sqrt{E}\,.\,dp.$$

If dp is positive, the change ds will be taken in the positive direction. Therefore $ds=\sqrt{E}\,.\,dp$,

$$\cos\alpha=\frac{1}{\sqrt{E}}\cdot\frac{\partial x}{\partial p},\quad \cos\beta=\frac{1}{\sqrt{E}}\cdot\frac{\partial y}{\partial p},\quad \cos\gamma=\frac{1}{\sqrt{E}}\cdot\frac{\partial z}{\partial p},$$

In like manner, along the line $p=$Constant, if $\cos\alpha'$, $\cos\beta'$, $\cos\gamma'$ are the direction cosines, we obtain

$$\cos\alpha'=\frac{1}{\sqrt{G}}\cdot\frac{\partial x}{\partial q},\quad \cos\beta'=\frac{1}{\sqrt{G}}\cdot\frac{\partial y}{\partial q},\quad \cos\gamma'=\frac{1}{\sqrt{G}}\cdot\frac{\partial z}{\partial q}.$$

And since

$$\cos\omega=\cos\alpha\cos\alpha'+\cos\beta\cos\beta'+\cos\gamma\cos\gamma',$$

$$\cos\omega=\frac{F}{\sqrt{EG}}.$$

From this follows

$$\sin\omega=\frac{\sqrt{(EG-F^2)}}{\sqrt{EG}}.$$

And the area of the quadrilateral formed by the lines $p, p+dp, q, q+dq$ is

$$d\sigma=\sqrt{(EG-F^2)}\,.\,dp\,.\,dq.$$

21. Art. 21, p. 33, l. 12. In the original, in Liouville's reprint, in the two French translations, and in Böklen's translation, the next to the last formula of this article is written

$$E\beta\delta-F(\alpha\delta+\beta\gamma)+G\alpha\gamma=\frac{EG-F'^2}{E'G'-F'^2}\cdot F'$$

The proper correction in sign has been made in Gauss, Werke, vol. IV, and in Wangerin's translation.

23. Art. 23, p. 35, l. 13 fr. bot. In the Latin texts and in Roger's and Böklen's translations this formula has a minus sign on the right hand side. The correction in sign has been made in Abadie's and Wangerin's translations.

22. Art. 23, p. 35. The figure below represents the lines and angles mentioned in this and the following articles:

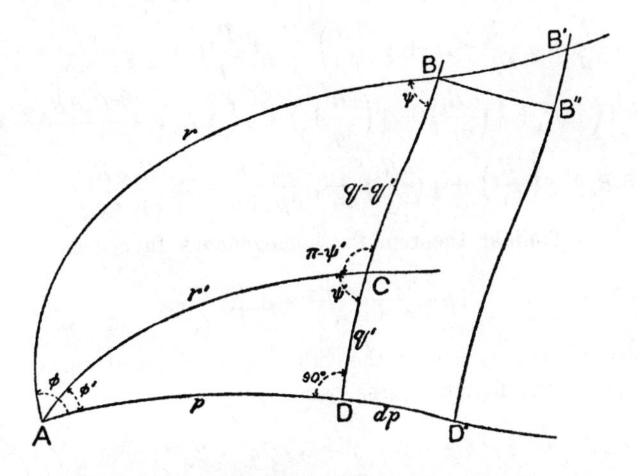

24. Art. 24, p. 36. Derivation of formula [1].

$$\text{Let } r^2 = p^2 + q^2 + R_3 + R_4 + R_5 + R_6 + \text{etc.}$$

where R_3 is the aggregate of all the terms of the third degree in p and q, R_4 of all the terms of the fourth degree, etc. Then by differentiating, squaring, and omitting terms above the sixth degree, we obtain

$$\left(\frac{\partial(r^2)}{\partial p}\right)^2 = 4p^2 + \left(\frac{\partial R_3}{\partial p}\right)^2 + \left(\frac{\partial R_4}{\partial p}\right)^2 + 4p\left(\frac{\partial R_3}{\partial p}\right) + 4p\frac{\partial R_4}{\partial p}$$
$$+ 4p\frac{\partial R_5}{\partial p} + 4p\frac{\partial R_6}{\partial p} + 2\frac{\partial R_3}{\partial p}\frac{\partial R_4}{\partial p} + 2\frac{\partial R_3}{\partial p}\frac{\partial R_5}{\partial p},$$

and

$$\left(\frac{\partial(r^2)}{\partial q}\right)^2 = 4q^2 + \left(\frac{\partial R_3}{\partial q}\right)^2 + \left(\frac{\partial R_4}{\partial q}\right)^2 + 4q\frac{\partial R_3}{\partial q} + 4q\frac{\partial R_4}{\partial q}$$
$$+ 4q\frac{\partial R_5}{\partial q} + 4q\frac{\partial R_6}{\partial q} + 2\frac{\partial R_3}{\partial q}\frac{\partial R_4}{\partial q} + 2\frac{\partial R_3}{\partial q}\frac{\partial R_5}{\partial q}.$$

Hence we have

$$\left(\frac{\partial(r^2)}{\partial p}\right)^2+\left(\frac{\partial(r^2)}{\partial q}\right)^2-4\,r^2$$

$$=4\left(p\frac{\partial R_3}{\partial p}+q\frac{\partial R_3}{\partial q}-R_3\right)+4\left(p\frac{\partial R_4}{\partial p}+q\frac{\partial R_4}{\partial q}-R_4+\tfrac{1}{4}\left(\frac{\partial R_3}{\partial p}\right)^2+\tfrac{1}{4}\left(\frac{\partial R_3}{\partial q}\right)^2\right)$$

$$+4\left(p\frac{\partial R_5}{\partial p}+q\frac{\partial R_5}{\partial q}-R_5+\tfrac{1}{2}\frac{\partial R_3}{\partial p}\frac{\partial R_4}{\partial p}+\tfrac{1}{2}\frac{\partial R_3}{\partial q}\frac{\partial R_4}{\partial q}\right)$$

$$+4\left(p\frac{\partial R_6}{\partial p}+q\frac{\partial R_6}{\partial q}-R_6+\tfrac{1}{4}\left(\frac{\partial R_4}{\partial p}\right)^2+\tfrac{1}{4}\left(\frac{\partial R_4}{\partial q}\right)^2+\tfrac{1}{2}\frac{\partial R_3}{\partial p}\frac{\partial R_5}{\partial p}+\tfrac{1}{2}\frac{\partial R_3}{\partial q}\frac{\partial R_5}{\partial q}\right)$$

$$=8\,R_3+4\left(3\,R_4+\tfrac{1}{4}\left(\frac{\partial R_3}{\partial p}\right)^2+\tfrac{1}{4}\left(\frac{\partial R_3}{\partial q}\right)^2\right)+4\left(4\,R_5+\tfrac{1}{2}\frac{\partial R_3}{\partial p}\frac{\partial R_4}{\partial p}+\tfrac{1}{2}\frac{\partial R_3}{\partial q}\frac{\partial R_4}{\partial q}\right)$$

$$+4\left(5\,R_6+\tfrac{1}{4}\left(\frac{\partial R_4}{\partial p}\right)^2+\tfrac{1}{4}\left(\frac{\partial R_4}{\partial q}\right)^2+\tfrac{1}{2}\frac{\partial R_3}{\partial p}\frac{\partial R_5}{\partial p}+\tfrac{1}{2}\frac{\partial R_3}{\partial q}\frac{\partial R_5}{\partial q}\right),$$

since, according to a familiar theorem for homogeneous functions,

$$p\frac{\partial R_3}{\partial p}+q\frac{\partial R_3}{\partial q}=3\,R_3,\ \text{etc.}$$

By dividing unity by the square of the value of n, given at the end of Art. 23, and omitting terms above the fourth degree, we have

$$1-\frac{1}{n^2}=2f^\circ q^2+2f'pq^2+2g^\circ q^3-3f^{\circ 2}q^4+2f''p^2q^2+2g'pq^3+2h^\circ q^4.$$

This, multiplied by the last equation but one of the preceding page, on rejecting terms above the sixth degree, becomes

$$\begin{aligned}\left(1-\frac{1}{n^2}\right)\left(\frac{\partial(r^2)}{\partial p}\right)^2=\;&8f^\circ p^2q^2+8f'p^3q^2 \quad -12f^{\circ 2}p^2q^4+8\,h^\circ p^2q^4\\&+8\,g^\circ p^2q^3 \quad +8f''p^4q^2 \quad +2f^\circ q^2\left(\frac{\partial R_3}{\partial p}\right)^2\\&+8f^\circ pq^2\frac{\partial R_3}{\partial p}+8\,g'p^3q^3 \quad +8f'p^2q^2\frac{\partial R_3}{\partial p}+8\,g^\circ pq^3\frac{\partial R_3}{\partial p}\\&+8f^\circ pq^2\frac{\partial R_4}{\partial p}.\end{aligned}$$

Therefore, since from the fifth equation of Art. 24:

$$\left(\frac{\partial(r^2)}{\partial p}\right)^2+\left(\frac{\partial(r^2)}{\partial q}\right)^2-4\,r^2=\left(1-\frac{1}{n^2}\right)\left(\frac{\partial(r^2)}{\partial p}\right)^2,$$

we have

$$\begin{aligned}
8\,R_3 &+ 4\left(3\,R_4 + \tfrac{1}{4}\left(\frac{\partial R_3}{\partial p}\right)^2 + \tfrac{1}{4}\left(\frac{\partial R_3}{\partial q}\right)^2\right) + 4\left(4\,R_5 + \tfrac{1}{2}\frac{\partial R_3}{\partial p}\frac{\partial R_4}{\partial p} + \tfrac{1}{2}\frac{\partial R_3}{\partial q}\frac{\partial R_4}{\partial q}\right)\\
&+ 4\left(5\,R_6 + \tfrac{1}{4}\left(\frac{\partial R_4}{\partial p}\right)^2 + \tfrac{1}{4}\left(\frac{\partial R_4}{\partial q}\right)^2 + \tfrac{1}{2}\frac{\partial R_3}{\partial p}\frac{\partial R_5}{\partial p} + \tfrac{1}{2}\frac{\partial R_3}{\partial q}\frac{\partial R_5}{\partial q}\right)\\
&\quad = 8\,f^\circ p^2q^2 + 8\,f'\,p^3q^2 + 8\,g^\circ p^2q^3 + 8\,f^\circ p\,q^2\frac{\partial R_3}{\partial p} - 12\,f^{\circ 2}p^2q^4 + 8\,f''p^4q^2\\
&\qquad + 8\,g'\,p^3q^3 + 8\,h^\circ p^2q^4 + 2\,f^\circ q^2\left(\frac{\partial R_3}{\partial p}\right)^2 + 8\,f'\,p^2q^2\frac{\partial R_3}{\partial p} + 8\,f^\circ p\,q^2\frac{\partial R_4}{\partial p} + 8\,g^\circ p\,q^3\frac{\partial R_3}{\partial p}.
\end{aligned}$$

Whence, by the method of undetermined coefficients, we find

$$\begin{aligned}
&R_3 = 0, \quad R_4 = \tfrac{2}{3}f^\circ p^2q^2, \quad R_5 = \tfrac{1}{2}f'\,p^3q^2 + \tfrac{1}{2}g^\circ\,p^2q^3,\\
&R_6 = (\tfrac{2}{5}f'' - \tfrac{4}{45}f^{\circ 2})\,p^4q^2 + \tfrac{2}{5}g'\,p^3q^3 + (\tfrac{2}{5}\,h^\circ - \tfrac{7}{45}f^{\circ 2})\,p^2q^4.
\end{aligned}$$

And therefore we have

[1]
$$\begin{aligned}
r^2 = p^2 &+ \tfrac{2}{3}f^\circ p^2q^2 + \tfrac{1}{2}f'\,p^3q^2 + (\tfrac{2}{5}f'' - \tfrac{4}{45}f^{\circ 2})\,p^4q^2 + \text{etc.}\\
+ q^2 &\qquad\qquad + \tfrac{1}{2}g^\circ p^2q^3 + \tfrac{2}{5}g'\,p^3q^3\\
&\qquad\qquad + (\tfrac{2}{5}\,h^\circ - \tfrac{7}{45}f^{\circ 2})\,p^2q^4.
\end{aligned}$$

This method for deriving formula [1] is taken from Wangerin.

25. Art. 24, p. 36. Derivation of formula [2].

By taking one-half the reciprocal of the series for n given in Art. 23, p. 36, we obtain

$$\frac{1}{2n} = \tfrac{1}{2}\,[1 - f^\circ q^2 - f'\,p\,q^2 - g^\circ q^3 - f''\,p^2q^2 - g'\,p\,q^3 - (h^\circ - f^{\circ 2})\,q^4 - \text{etc.}].$$

And by differentiating formula [1] with respect to p, we obtain

$$\begin{aligned}
\frac{\partial(r^2)}{\partial p} = 2\,[p &+ \tfrac{2}{3}f^\circ p\,q^2 + \tfrac{3}{4}f'\,p^2q^2 + 2\,(\tfrac{2}{5}f'' - \tfrac{4}{45}f^{\circ 2})\,p^3q^2\\
&+ \tfrac{1}{2}g^\circ p\,q^3 + \tfrac{3}{5}g'\,p^2q^3\\
&+ (\tfrac{2}{5}\,h^\circ - \tfrac{7}{45}f^{\circ 2})\,p\,q^4 + \text{etc.}].
\end{aligned}$$

Therefore, since

$$r\sin\psi = \frac{1}{2n}\cdot\frac{\partial(r^2)}{\partial p},$$

we have, by multiplying together the two series above,

[2]
$$\begin{aligned}
r\sin\psi = p &- \tfrac{1}{3}f^\circ p\,q^2 - \tfrac{1}{4}f'\,p^2q^2 - (\tfrac{1}{5}f'' + \tfrac{8}{45}f^{\circ 2})\,p^3q^2 - \text{etc.}\\
&- \tfrac{1}{2}g^\circ p\,q^3 - \tfrac{2}{5}g'\,p^2q^3\\
&- (\tfrac{3}{5}\,h^\circ - \tfrac{8}{45}f^{\circ 2})\,p\,q^4.
\end{aligned}$$

26. Art. 24, p. 37. Derivation of formula [3].

By differentiating [1] on page 57 with respect to q, we find

$$\frac{\partial (r^2)}{\partial q} = 2\,[q + \tfrac{2}{3} f^\circ p^2 q + \tfrac{1}{2} f' p^3 q + (\tfrac{2}{5} f'' - \tfrac{4}{45} f^{\circ 2}) p^4 q + \tfrac{3}{4} g^\circ p^2 q^2 + \tfrac{3}{5} g' p^3 q^2 + (\tfrac{4}{5} h^\circ - \tfrac{14}{45} f^{\circ 2}) p^2 q^3 + \text{etc.}].$$

Therefore we have, since

$$r \cos \psi = \tfrac{1}{2} \frac{\partial (r^2)}{\partial q},$$

$$[3] \qquad r \cos \psi = q + \tfrac{2}{3} f^\circ p^2 q + \tfrac{1}{2} f' p^3 q + (\tfrac{2}{5} f'' - \tfrac{4}{45} f^{\circ 2}) p^4 q + \text{etc.} + \tfrac{3}{4} g^\circ p^2 q^2 + \tfrac{3}{5} g' p^3 q^2 + (\tfrac{4}{5} h^\circ - \tfrac{14}{45} f^{\circ 2}) p^2 q^3.$$

27. Art. 24, p. 37. Derivation of formula [4].

Since $r \cos \phi$ becomes equal to p for infinitely small values of p and q, the series for $r \cos \phi$ must begin with p. Hence we set

$$(1) \qquad r \cos \phi = p + R_2 + R_3 + R_4 + R_5 + \text{etc.}$$

Then, by differentiating, we obtain

$$(2) \qquad \frac{\partial (r \cos \phi)}{\partial p} = 1 + \frac{\partial R_2}{\partial p} + \frac{\partial R_3}{\partial p} + \frac{\partial R_4}{\partial p} + \frac{\partial R_5}{\partial p} + \text{etc.}$$

$$(3) \qquad \frac{\partial (r \cos \phi)}{\partial q} = \frac{\partial R_2}{\partial q} + \frac{\partial R_3}{\partial q} + \frac{\partial R_4}{\partial q} + \frac{\partial R_5}{\partial q} + \text{etc.}$$

By dividing [2] p. 57 by n on page 36, we obtain

$$(4) \qquad \frac{r \sin \psi}{n} = p - \tfrac{4}{3} f^\circ p q^2 - \tfrac{5}{4} f' p^2 q^2 - (\tfrac{6}{5} f'' + \tfrac{8}{45} f^{\circ 2}) p^3 q^2 - \text{etc.} - \tfrac{3}{2} g^\circ p q^3 - \tfrac{7}{5} g' p^2 q^3 - (\tfrac{8}{5} h^\circ - \tfrac{68}{45} f^{\circ 2}) p q^4.$$

Multiplying (2) by (4), we have

$$(5) \qquad \frac{r \sin \psi}{n} \cdot \frac{\partial (r \cos \phi)}{\partial p} = p + p \frac{\partial R_2}{\partial p} + p \frac{\partial R_3}{\partial p} + p \frac{\partial R_4}{\partial p} + p \frac{\partial R_5}{\partial p} - (\tfrac{6}{5} f'' + \tfrac{8}{45} f^{\circ 2}) p^3 q^2$$
$$- \tfrac{4}{3} f^\circ p q^2 - \tfrac{4}{3} f^\circ p q^2 \frac{\partial R_2}{\partial p} - \tfrac{4}{3} f^\circ p q^2 \frac{\partial R_3}{\partial p} - \tfrac{7}{5} g' p^2 q^3$$
$$- \tfrac{5}{4} f' p^2 q^2 - \tfrac{5}{4} f' p^2 q^2 \frac{\partial R_2}{\partial p} - (\tfrac{8}{5} h^\circ - \tfrac{68}{45} f^{\circ 2}) p q^4$$
$$- \tfrac{3}{2} g^\circ p q^3 - \tfrac{3}{2} g^\circ p q^3 \frac{\partial R_2}{\partial p} - \text{etc.}$$

Multiplying (3) by [3] p. 58, we have

$$(6)\quad r\cos\psi\cdot\frac{\partial(r\cos\phi)}{\partial q}=q\frac{\partial R_2}{\partial q}+q\frac{\partial R_3}{\partial q}+q\frac{\partial R_4}{\partial q}+q\frac{\partial R_5}{\partial q}+\tfrac{1}{2}f'p^3q\frac{\partial R_2}{\partial q}$$
$$+\tfrac{2}{3}f^\circ p^2q\frac{\partial R_2}{\partial q}+\tfrac{2}{3}f^\circ p^2q\frac{\partial R_3}{\partial q}+\tfrac{3}{4}g^\circ p^2q^2\frac{\partial R_2}{\partial q}+\text{etc.}$$

Since

$$\frac{r\sin\psi}{n}\cdot\frac{\partial(r\cos\phi)}{\partial p}+r\cos\psi\cdot\frac{\partial(r\cos\phi)}{\partial q}=r\cos\phi,$$

we have, by setting (1) equal to the sum of (5) and (6),

$$p+R_2+R_3+R_4+R_5+\text{etc.}$$
$$=p+p\frac{\partial R_2}{\partial p}+p\frac{\partial R_3}{\partial p}+p\frac{\partial R_4}{\partial p}+p\frac{\partial R_5}{\partial p}-(\tfrac{8}{5}h^\circ-\tfrac{68}{45}f^{\circ 2})pq^4$$
$$+q\frac{\partial R_2}{\partial q}-\tfrac{4}{3}f^\circ pq^2-\tfrac{4}{3}f^\circ pq^2\frac{\partial R_2}{\partial p}-\tfrac{4}{3}f^\circ pq^2\frac{\partial R_3}{\partial p}+q\frac{\partial R_5}{\partial q}$$
$$+q\frac{\partial R_3}{\partial q}-\tfrac{5}{4}f'p^2q^2-\tfrac{5}{4}f'p^2q^2\frac{\partial R_2}{\partial p}+\tfrac{2}{3}f^\circ p^2q\frac{\partial R_3}{\partial q}$$
$$-\tfrac{3}{2}g^\circ pq^3-\tfrac{3}{2}g^\circ pq^3\frac{\partial R_2}{\partial p}+\tfrac{1}{2}f'p^3q\frac{\partial R_2}{\partial q}$$
$$+q\frac{\partial R_4}{\partial q}-(\tfrac{6}{5}f''+\tfrac{8}{45}f^{\circ 2})p^3q^2+\tfrac{3}{4}g^\circ p^2q^2\frac{\partial R_2}{\partial q}$$
$$+\tfrac{2}{3}f^\circ p^2q\frac{\partial R_2}{\partial q}-\tfrac{7}{5}g'p^2q^3+\text{etc.},$$

from which we find

$$R_2=0,\qquad R_3=\tfrac{2}{3}f^\circ pq^2,\qquad R_4=\tfrac{5}{12}f'p^2q^2+\tfrac{1}{2}g^\circ pq^3,$$
$$R_5=\tfrac{7}{20}g'p^2q^3+(\tfrac{2}{5}h^\circ-\tfrac{7}{45}f^{\circ 2})pq^4+(\tfrac{3}{10}f''-\tfrac{8}{45}f^{\circ 2})p^3q^2.$$

Therefore we have finally

$$[4]\qquad r\cos\phi=p+\tfrac{2}{3}f^\circ pq^2+\tfrac{5}{12}f'p^2q^2+(\tfrac{3}{10}f''-\tfrac{8}{45}f^{\circ 2})p^3q^2+\text{etc.}$$
$$+\tfrac{1}{2}g^\circ pq^3+\tfrac{7}{20}g'p^2q^3$$
$$+(\tfrac{2}{5}h^\circ-\tfrac{7}{45}f^{\circ 2})pq^4.$$

28. Art. 24, p. 37. Derivation of formula [5].

Again, since $r\sin\phi$ becomes equal to q for infinitely small values of p and q, we set

$$(1)\qquad r\sin\phi=q+R_2+R_3+R_4+R_5+\text{etc.}$$

Then we have by differentiation

$$(2)\qquad \frac{\partial (r\sin\phi)}{\partial p}=\frac{\partial R_2}{\partial p}+\frac{\partial R_3}{\partial p}+\frac{\partial R_4}{\partial p}+\frac{\partial R_5}{\partial p}+\text{etc.}$$

$$(3)\qquad \frac{\partial (r\sin\phi)}{\partial q}=1+\frac{\partial R_2}{\partial q}+\frac{\partial R_3}{\partial q}+\frac{\partial R_4}{\partial q}+\frac{\partial R_5}{\partial q}+\text{etc.}$$

Multiplying (4) p. 58 by this (2), we obtain

$$(4)\quad \frac{r\sin\psi}{n}\cdot\frac{\partial (r\sin\phi)}{\partial p}=p\frac{\partial R_2}{\partial p}+p\frac{\partial R_3}{\partial p}+p\frac{\partial R_4}{\partial p}+p\frac{\partial R_5}{\partial p}-\tfrac{5}{4}f'p^2q^2\frac{\partial R_2}{\partial p}$$
$$-\tfrac{4}{3}f^\circ pq^2\frac{\partial R_2}{\partial p}-\tfrac{4}{3}f^\circ pq^2\frac{\partial R_3}{\partial p}-\tfrac{3}{2}g^\circ pq^3\frac{\partial R_2}{\partial p}-\text{etc.}$$

Likewise from (3) and [3] p. 58, we obtain

$$(5)\quad r\cos\psi\cdot\frac{\partial (r\sin\phi)}{\partial q}=q+q\frac{\partial R_2}{\partial q}+q\frac{\partial R_3}{\partial q}+q\frac{\partial R_4}{\partial q}+q\frac{\partial R_5}{\partial q}+(\tfrac{2}{5}f''-\tfrac{4}{45}f^{\circ 2})p^4q$$
$$+\tfrac{2}{3}f^\circ p^2q+\tfrac{2}{3}f^\circ p^2q\frac{\partial R_2}{\partial q}+\tfrac{2}{3}f^\circ p^2q\frac{\partial R_3}{\partial q}+\tfrac{3}{5}g'p^3q^2$$
$$+\tfrac{1}{2}f'p^3q+\tfrac{1}{2}f'p^3q\frac{\partial R_2}{\partial q}+(\tfrac{4}{5}h^\circ-\tfrac{14}{45}f^{\circ 2})p^2q^3$$
$$+\tfrac{3}{4}g^\circ p^2q^2+\tfrac{3}{4}g^\circ p^2q^2\frac{\partial R_2}{\partial q}+\text{etc.}$$

Since

$$\frac{r\sin\psi}{n}\cdot\frac{\partial (r\sin\phi)}{\partial p}+r\cos\psi\cdot\frac{\partial (r\sin\phi)}{\partial q}=r\sin\phi,$$

by setting (1) equal to the sum of (4) and (5), we have

$$q+R_2+R_3+R_4+R_5+\text{etc.}$$
$$=q+p\frac{\partial R_2}{\partial p}+p\frac{\partial R_3}{\partial p}+p\frac{\partial R_4}{\partial p}+\tfrac{1}{2}f'p^3q+p\frac{\partial R_5}{\partial p}+\tfrac{1}{2}f'p^3q\frac{\partial R_2}{\partial q}+q\frac{\partial R_5}{\partial q}+\text{etc.}$$
$$+q\frac{\partial R_2}{\partial q}+q\frac{\partial R_3}{\partial q}+q\frac{\partial R_4}{\partial q}+\tfrac{3}{4}g^\circ p^2q^2-\tfrac{4}{3}f^\circ pq^2\frac{\partial R_3}{\partial p}+\tfrac{3}{4}g^\circ p^2q^2\frac{\partial R_2}{\partial q}+\tfrac{2}{3}f^\circ p^2q\frac{\partial R_3}{\partial q}$$
$$+\tfrac{2}{3}f^\circ p^2q-\tfrac{4}{3}f^\circ pq^2\frac{\partial R_2}{\partial p}-\tfrac{5}{4}f'p^2q^2\frac{\partial R_2}{\partial p}+(\tfrac{2}{5}f''-\tfrac{4}{45}f^{\circ 2})p^4q$$
$$+\tfrac{2}{3}f^\circ p^2q\frac{\partial R_2}{\partial q}-\tfrac{3}{2}g^\circ pq^3\frac{\partial R_2}{\partial p}+\tfrac{3}{5}g'p^3q^2+(\tfrac{4}{5}h^\circ-\tfrac{14}{45}f^{\circ 2})p^2q^3.$$

from which we find

$$R_2=0,\quad R_3=-\tfrac{1}{3}f^\circ p^2 q,\qquad R_4=-\tfrac{1}{6}f' p^3 q-\tfrac{1}{4}g^\circ p^2 q^2,$$
$$R_5=-(\tfrac{1}{10}f''-\tfrac{7}{90}f^{\circ 2})p^4 q-\tfrac{3}{20}g' p^3 q^2-(\tfrac{1}{5}h^\circ+\tfrac{13}{90}f^{\circ 2})p^2 q^3.$$

Therefore, substituting these values in (1), we have

$$[5]\qquad \begin{aligned} r\sin\phi = q-\tfrac{1}{3}f^\circ p^2 q-\tfrac{1}{6}f' p^3 q\ &-(\tfrac{1}{10}f''-\tfrac{7}{90}f^{\circ 2})p^4 q-\text{etc.}\\ -\tfrac{1}{4}g^\circ p^2 q^2-\ &\tfrac{3}{20}g' p^3 q^2\\ &-(\tfrac{1}{5}h^\circ+\tfrac{13}{90}f^{\circ 2})p^2 q^3.\end{aligned}$$

Art. 24, p. 38. Derivation of formula [6].

Differentiating n on page 36 with respect to q, we obtain

$$(1)\qquad \begin{aligned}\frac{\partial n}{\partial q}=2f^\circ q+2f' pq+2f'' p^2 q &+ \text{etc.}\\ 3g^\circ q^2+3g' pq^2 &+ \text{etc.}\\ +4h^\circ q^3 &+ \text{etc., etc.}\end{aligned}$$

and hence, multiplying this series by (4) on page 58, we find

$$(2)\qquad \begin{aligned}\frac{r\sin\psi}{n}\cdot\frac{\partial n}{\partial q}=2f^\circ pq+2f' p^2 q+2f'' p^3 q+3g' p^2 q^2+\text{etc.}\\ +3g^\circ pq^2+(4h^\circ-\tfrac{8}{3}f^{\circ 2})pq^3.\end{aligned}$$

For infinitely small values of r, $\psi+\phi=\frac{\pi}{2}$, as is evident from the figure on page 55. Hence we set

$$\psi+\phi=\frac{\pi}{2}+R_1+R_2+R_3+R_4+\text{etc.}$$

Then we shall have, by differentiation,

$$(3)\qquad \frac{\partial(\psi+\phi)}{\partial p}=\frac{\partial R_1}{\partial p}+\frac{\partial R_2}{\partial p}+\frac{\partial R_3}{\partial p}+\frac{\partial R_4}{\partial p}+\text{etc.}$$

$$(4)\qquad \frac{\partial(\psi+\phi)}{\partial q}=\frac{\partial R_1}{\partial q}+\frac{\partial R_2}{\partial q}+\frac{\partial R_3}{\partial q}+\frac{\partial R_4}{\partial q}+\text{etc.}$$

Therefore, multiplying (4) on page 58 by (3), we find

$$(5)\qquad \begin{aligned}\frac{r\sin\psi}{n}\cdot\frac{\partial(\psi+\phi)}{\partial p}=p\frac{\partial R_1}{\partial p}+p\frac{\partial R_2}{\partial p}+p\frac{\partial R_3}{\partial p}\qquad &+p\frac{\partial R_4}{\partial p}+\text{etc.}\\ -\tfrac{4}{3}f^\circ pq^2\frac{\partial R_1}{\partial p}&-\tfrac{4}{3}f^\circ pq^2\frac{\partial R_2}{\partial p}\\ &-\tfrac{5}{4}f' p^2 q^2\frac{\partial R_1}{\partial p}\\ &-\tfrac{3}{2}g^\circ pq^3\frac{\partial R_1}{\partial p},\end{aligned}$$

and, multiplying [3] on page 58 by (4), we find

$$(6)\qquad r\cos\psi\cdot\frac{\partial(\psi+\phi)}{\partial q}=q\frac{\partial R_1}{\partial q}+q\frac{\partial R_2}{\partial q}+q\frac{\partial R_3}{\partial q}+q\frac{\partial R_4}{\partial q}+\text{etc.}$$
$$+\tfrac{2}{3}f^\circ p^2 q\frac{\partial R_1}{\partial q}+\tfrac{2}{3}f^\circ p^2 q\frac{\partial R_2}{\partial q}$$
$$+\tfrac{1}{2}f' p^3 q\frac{\partial R_1}{\partial q}$$
$$+\tfrac{3}{4}g^\circ p^2 q^2\frac{\partial R_1}{\partial q}.$$

And since

$$\frac{r\sin\psi}{n}\cdot\frac{\partial n}{\partial q}+\frac{r\sin\psi}{n}\cdot\frac{\partial(\psi+\phi)}{\partial p}+r\cos\psi\cdot\frac{\partial(\psi+\phi)}{\partial q}=0,$$

we shall have, by adding (2), (5), and (6),

$$0=p\frac{\partial R_1}{\partial p}+2f^\circ pq+2f'p^2q+2f''p^3q-\tfrac{3}{2}g^\circ pq^3\frac{\partial R_1}{\partial p}$$
$$q\frac{\partial R_1}{\partial q}+p\frac{\partial R_2}{\partial p}+3g^\circ pq^2+3g'p^2q^2+q\frac{\partial R_4}{\partial q}$$
$$+q\frac{\partial R_2}{\partial q}+p\frac{\partial R_3}{\partial p}+(4h^\circ-\tfrac{8}{3}f^{\circ 2})pq^3+\tfrac{2}{3}f^\circ p^2q\frac{\partial R_2}{\partial q}$$
$$-\tfrac{4}{3}f^\circ pq^2\frac{\partial R_1}{\partial p}+p\frac{\partial R_4}{\partial p}+\tfrac{1}{2}f'p^3q\frac{\partial R_1}{\partial q}$$
$$+q\frac{\partial R_3}{\partial q}-\tfrac{4}{3}f^\circ pq^2\frac{\partial R_2}{\partial p}+\tfrac{3}{4}g^\circ p^2q^2\frac{\partial R_1}{\partial q}$$
$$+\tfrac{2}{3}f^\circ p^2q\frac{\partial R_1}{\partial q}-\tfrac{5}{4}f^\circ p^2q^2\frac{\partial R_1}{\partial p}+\text{etc.}$$

From this equation we find

$$R_1=0,\quad R_2=-f^\circ pq,\quad R_3=-\tfrac{2}{3}f'p^2q-g^\circ pq^2,$$
$$R_4=-(\tfrac{1}{2}f''-\tfrac{1}{6}f^{\circ 2})p^3q-\tfrac{3}{4}g'p^2q^2-(h^\circ-\tfrac{1}{3}f^{\circ 2})pq^3.$$

Therefore we have finally

$$[6]\qquad \psi+\phi=\frac{\pi}{2}-f^\circ pq-\tfrac{2}{3}f'p^2q-(\tfrac{1}{2}f''-\tfrac{1}{6}f^{\circ 2})p^3q-\text{etc.}$$
$$-g^\circ pq^2-\tfrac{3}{4}g'p^2q^2$$
$$-(h^\circ-\tfrac{1}{3}f^{\circ 2})pq^3.$$

29. Art. 24, p. 38, l. 19. The differential equation from which formula [7] follows is derived in the following manner. In the figure on page 55, prolong AD to D', making $DD' = dp$, through D' perpendicular to AD' draw a geodesic line, which will cut AB in B'. Finally, take $D'B'' = DB$, so that BB'' is perpendicular to $B'D'$. Then, if by ABD we mean the area of the triangle ABD,

$$\frac{\partial S}{\partial r} = \lim \frac{AB'D' - ABD}{BB'} = \lim \frac{BDD'B'}{BB'} = \lim \frac{BDD'B''}{DD'} \cdot \lim \frac{DD'}{BB'},$$

since the surface $BDD'B''$ differs from $BDD'B'$ only by an infinitesimal of the second order. And since

$$BDD'B'' = dp \,.\int n\,dq, \text{ or } \lim \frac{BDD'B''}{DD'} = \int n\,dq,$$

and since, further,

$$\lim \frac{DD'}{BB'} = \frac{\partial p}{\partial r},$$

consequently

$$\frac{\partial S}{\partial r} = \frac{\partial p}{\partial r} \cdot \int n\,dq.$$

Therefore also

$$\frac{\partial S}{\partial p} \cdot \frac{\partial p}{\partial r} + \frac{\partial S}{\partial q} \cdot \frac{\partial q}{\partial r} = \frac{\partial p}{\partial r} \cdot \int n\,dq.$$

Finally, from the values for $\frac{\partial r}{\partial p}, \frac{\partial r}{\partial q}$ given at the beginning of Art. 24, p. 36, we have

$$\frac{\partial p}{\partial r} = \frac{1}{n} \sin \psi, \quad \frac{\partial q}{\partial r} = \cos \psi,$$

so that we have

$$\frac{\partial S}{\partial p} \cdot \frac{\sin \psi}{n} + \frac{\partial S}{\partial q} \cdot \cos \psi = \frac{\sin \psi}{n} \cdot \int n\,dq.$$

[Wangerin.]

30. Art. 24, p. 38. Derivation of formula [7].

For infinitely small values of p and q, the area of the triangle ABC becomes equal to $\frac{1}{2}pq$. The series for this area, which is denoted by S, must therefore begin with $\frac{1}{2}pq$, or R_2. Hence we put

$$S = R_2 + R_3 + R_4 + R_5 + R_6 + \text{etc.}$$

By differentiating, we obtain

$$\frac{\partial S}{\partial p} = \frac{\partial R_2}{\partial p} + \frac{\partial R_3}{\partial p} + \frac{\partial R_4}{\partial p} + \frac{\partial R_5}{\partial p} + \frac{\partial R_6}{\partial p} + \text{etc.}, \tag{1}$$

$$\frac{\partial S}{\partial q} = \frac{\partial R_2}{\partial q} + \frac{\partial R_3}{\partial q} + \frac{\partial R_4}{\partial q} + \frac{\partial R_5}{\partial q} + \frac{\partial R_6}{\partial q} + \text{etc.}, \tag{2}$$

and therefore, by multiplying (4) on page 58 by (1), we obtain

$$\begin{aligned}
\frac{r \sin \psi}{n} \cdot \frac{\partial S}{\partial p} = {} & p\frac{\partial R_2}{\partial p} + p\frac{\partial R_3}{\partial p} + p\frac{\partial R_4}{\partial p} + p\frac{\partial R_5}{\partial p} + p\frac{\partial R_6}{\partial p} + \text{etc.} \\
& - \tfrac{4}{3} f^\circ p\, q^2 \frac{\partial R_2}{\partial p} - \tfrac{4}{3} f^\circ p\, q^2 \frac{\partial R_3}{\partial p} - \tfrac{4}{3} f^\circ p\, q^2 \frac{\partial R_4}{\partial p} \\
& - \tfrac{5}{4} f' p^2 q^2 \frac{\partial R_2}{\partial p} - \tfrac{5}{4} f' p^2 q^2 \frac{\partial R_3}{\partial p} \\
& - \tfrac{3}{2} g^\circ p\, q^3 \frac{\partial R_2}{\partial p} - \tfrac{3}{2} g^\circ p\, q^3 \frac{\partial R_3}{\partial p} \\
& - \left(\tfrac{6}{5} f'' + \tfrac{8}{45} f^{\circ 2}\right) p^3 q^2 \frac{\partial R_2}{\partial p} \\
& - \tfrac{7}{5} g' p^2 q^3 \frac{\partial R_2}{\partial p} \\
& - \left(\tfrac{8}{5} h^\circ - \tfrac{68}{45} f^{\circ 2}\right) p\, q^4 \frac{\partial R_2}{\partial p},
\end{aligned} \tag{3}$$

and multiplying [3] on page 58 by (2), we obtain

$$\begin{aligned}
r \cos \psi \cdot \frac{\partial S}{\partial q} = {} & q\frac{\partial R_2}{\partial q} + q\frac{\partial R_3}{\partial q} + q\frac{\partial R_4}{\partial q} + q\frac{\partial R_5}{\partial q} + q\frac{\partial R_6}{\partial q} + \text{etc.} \\
& + \tfrac{2}{3} f^\circ p^2 q \frac{\partial R_2}{\partial q} + \tfrac{2}{3} f^\circ p^2 q \frac{\partial R_3}{\partial q} + \tfrac{2}{3} f^\circ p^2 q \frac{\partial R_4}{\partial q} \\
& + \tfrac{1}{2} f' p^3 q \frac{\partial R_2}{\partial q} + \tfrac{1}{2} f' p^3 q \frac{\partial R_3}{\partial q} \\
& + \tfrac{3}{4} g^\circ p^2 q^2 \frac{\partial R_2}{\partial q} + \tfrac{3}{4} g^\circ p^2 q^2 \frac{\partial R_3}{\partial q} \\
& + \left(\tfrac{2}{5} f'' - \tfrac{4}{45} f^{\circ 2}\right) p^4 q \frac{\partial R_2}{\partial q} \\
& + \tfrac{3}{5} g' p^3 q^2 \frac{\partial R_2}{\partial q} \\
& + \left(\tfrac{4}{5} h^\circ - \tfrac{14}{45} f^{\circ 2}\right) p^2 q^3 \frac{\partial R_2}{\partial q}.
\end{aligned} \tag{4}$$

Integrating n on page 36 with respect to q, we find

$$(5)\qquad \int n\,dq = q + \tfrac{1}{3}f^\circ q^3 + \tfrac{1}{3}f' p q^3 + \tfrac{1}{3}f'' p^2 q^3 + \text{etc.} \\ + \tfrac{1}{4}g^\circ q^4 + \tfrac{1}{4}g' p q^4 + \text{etc.} \\ + \tfrac{1}{5}h^\circ q^5 + \text{etc. etc.}$$

Multiplying (4) on page 58 by (5), we find

$$(6)\qquad \frac{r\sin\psi}{n}\cdot\int n\,dq = pq - f^\circ pq^3 - \tfrac{11}{12}f' p^2 q^3 - (\tfrac{13}{15}f'' + \tfrac{8}{45}f^{\circ 2})p^3q^3 - \text{etc.} \\ - \tfrac{5}{4}g^\circ pq^4 - \tfrac{23}{20}g' p^2 q^4 \\ - (\tfrac{7}{5}h^\circ - \tfrac{16}{15}f^{\circ 2})pq^5.$$

Since

$$\frac{r\sin\psi}{n}\cdot\frac{\partial S}{\partial p} + r\cos\psi\cdot\frac{\partial S}{\partial q} = \frac{r\sin\psi}{n}\cdot\int n\,dq,$$

we obtain, by setting (6) equal to the sum of (3) and (4),

$$pq \quad - f^\circ pq^3 \quad - \tfrac{11}{12}f'p^2q^3 \quad - (\tfrac{13}{15}f'' + \tfrac{8}{45}f^{\circ 2})p^3q^3 - \text{etc.} \\ - \tfrac{5}{4}g^\circ pq^4 \quad - \tfrac{23}{20}g'p^2q^4 \\ - (\tfrac{7}{5}h^\circ - \tfrac{16}{15}f^{\circ 2})pq^5$$

$$= p\frac{\partial R_2}{\partial p} + p\frac{\partial R_3}{\partial p} + p\frac{\partial R_4}{\partial p} \qquad + p\frac{\partial R_5}{\partial p} + q\frac{\partial R_5}{\partial q} + p\frac{\partial R_6}{\partial p} \qquad + \tfrac{3}{4}g^\circ p^2q^2\frac{\partial R_3}{\partial q} + \text{etc.} \\ + q\frac{\partial R_2}{\partial q} + q\frac{\partial R_3}{\partial q} + q\frac{\partial R_4}{\partial q} \qquad - \tfrac{4}{3}f^\circ pq^2\frac{\partial R_3}{\partial p} + q\frac{\partial R_6}{\partial q} \qquad - (\tfrac{6}{5}f'' + \tfrac{8}{45}f^{\circ 2})p^3q^2\frac{\partial R_2}{\partial p} \\ - \tfrac{4}{3}f^\circ pq^2\frac{\partial R_2}{\partial p} + \tfrac{2}{3}f^\circ p^2q\frac{\partial R_3}{\partial q} - \tfrac{4}{3}f^\circ pq^2\frac{\partial R_4}{\partial p} + (\tfrac{2}{5}f'' - \tfrac{4}{45}f^{\circ 2})p^4q\frac{\partial R_2}{\partial q} \\ + \tfrac{2}{3}f^\circ p^2q\frac{\partial R_2}{\partial q} - \tfrac{5}{4}f'p^2q^2\frac{\partial R_2}{\partial p} + \tfrac{2}{3}f^\circ p^2q\frac{\partial R_4}{\partial q} - \tfrac{7}{5}g'p^2q^3\frac{\partial R_2}{\partial p} \\ + \tfrac{1}{2}f'p^3q\frac{\partial R_2}{\partial q} - \tfrac{5}{4}f'p^2q^2\frac{\partial R_3}{\partial p} + \tfrac{3}{5}g'p^3q^2\frac{\partial R_2}{\partial q} \\ - \tfrac{3}{2}g^\circ pq^3\frac{\partial R_2}{\partial p} + \tfrac{1}{2}f'p^3q\frac{\partial R_3}{\partial q} - (\tfrac{8}{5}h^\circ - \tfrac{68}{45}f^{\circ 2})pq^4\frac{\partial R_2}{\partial p} \\ + \tfrac{3}{4}g^\circ p^2q^2\frac{\partial R_2}{\partial q} - \tfrac{3}{2}g^\circ pq^3\frac{\partial R_3}{\partial p} + (\tfrac{4}{5}h^\circ - \tfrac{14}{45}f^{\circ 2})p^2q^3\frac{\partial R_2}{\partial q}.$$

From this equation we find

$$R_2 = \tfrac{1}{2}pq, \quad R_3 = 0, \quad R_4 = -\tfrac{1}{12}f^\circ pq^3 - \tfrac{1}{12}f^\circ p^3q, \\ R_5 = -\tfrac{1}{20}f'p^4q - \tfrac{3}{40}g^\circ p^3q^2 - \tfrac{7}{120}f'p^2q^3 - \tfrac{1}{10}g^\circ pq^4, \\ R_6 = -(\tfrac{1}{10}h^\circ - \tfrac{1}{30}f^{\circ 2})pq^5 - (\tfrac{1}{15}h^\circ + \tfrac{2}{45}f'' + \tfrac{1}{60}f^{\circ 2})p^3q^3 \\ - \tfrac{3}{40}g'p^2q^4 - (\tfrac{1}{30}f'' - \tfrac{1}{60}f^{\circ 2})p^5q - \tfrac{1}{20}g'p^4q^2.$$

Therefore we have

[7] $$\begin{aligned} S = \tfrac{1}{2} p q - \tfrac{1}{12} f^\circ p q^3 - \tfrac{1}{20} f' p^4 q & - (\tfrac{1}{30} f'' - \tfrac{1}{60} f^{\circ 2}) p^5 q - \text{etc.} \\ - \tfrac{1}{12} f^\circ p^3 q - \tfrac{3}{40} g^\circ p^3 q^2 & - \tfrac{1}{20} g' p^4 q^2 \\ - \tfrac{7}{120} f' p^2 q^3 & - (\tfrac{1}{15} h^\circ + \tfrac{2}{45} f'' + \tfrac{1}{60} f^{\circ 2}) p^3 q^3 \\ - \tfrac{1}{10} g^\circ p q^4 & - \tfrac{3}{40} g' p^2 q^4 \\ & - (\tfrac{1}{10} h^\circ - \tfrac{1}{30} f^{\circ 2}) p q^5. \end{aligned}$$

31. Art. 25, p. 39, l. 17. $3 p^2 + 4 q^2 + 4 q q' + 4 q'^2$ is replaced by $3 p^2 + 4 q^2 + 4 q'^2$. This error appears in all the reprints and translations (except Wangerin's).

32. Art. 25, p. 40, l. 8. $3 p^2 - 2 q^2 + q q' + 4 q q'$ is replaced by $3 p^2 - 2 q^2 + q q' + 4 q'^2$. This correction is noted in all the translations, and in Liouville's reprint.

33. Art. 25, p. 40. Derivation of formulæ [8], [9], [10].

By priming the q's in [7] we obtain at once a series for S'. Then, since $\sigma = S - S'$, we have

$$\begin{aligned} \sigma = \tfrac{1}{2} p (q - q') - \tfrac{1}{12} f^\circ p^3 (q - q') - \tfrac{1}{20} f' p^4 (q - q') - \tfrac{3}{40} g^\circ p^3 (q^2 - q'^2) \\ - \tfrac{1}{12} f^\circ p (q^3 - q'^3) - \tfrac{7}{120} f' p^2 (q^3 - q'^3) - \tfrac{1}{10} g^\circ p (q^4 - q'^4), \end{aligned}$$

correct to terms of the sixth degree.

This expression may be written as follows:

$$\begin{aligned} \sigma = \tfrac{1}{2} p (q - q') \big(1 & - \tfrac{1}{6} f^\circ (p^2 + q^2 + q q' + q'^2) \\ & - \tfrac{1}{60} f' p (6 p^2 + 7 q^2 + 7 q q' + 7 q'^2) \\ & - \tfrac{1}{20} g^\circ (q + q') (3 p^2 + 4 q^2 + 4 q'^2)\big), \end{aligned}$$

or, after factoring,

(1) $$\begin{aligned} \sigma = \tfrac{1}{2} p (q - q') \big(1 - \tfrac{1}{3} f^\circ q^2 - \tfrac{1}{4} f' p q^2 - \tfrac{1}{2} g^\circ q^3\big) \big(1 - \tfrac{1}{6} f^\circ (p^2 - q^2 + q q' + q'^2) \\ - \tfrac{1}{60} f' p (6 p^2 - 8 q^2 + 7 q q' + 7 q'^2) - \tfrac{1}{20} g^\circ (3 p^2 q + 3 p^2 q' - 6 q^3 + 4 q^2 q' + 4 q q'^2 + 4 q'^3)\big). \end{aligned}$$

The last factor on the right in (1) can be written thus:

$$\begin{aligned} \big(1 & - \tfrac{2}{120} f^\circ (4 p^2) - \tfrac{2}{120} f^\circ (3 p^2) - \tfrac{2}{120} f' p (6 q q') - \tfrac{2}{120} f^\circ (3 p^2) - \tfrac{2}{120} f' p (q q') \\ & + \tfrac{2}{120} f^\circ (2 q^2) + \tfrac{2}{120} f^\circ (6 q^2) - \tfrac{2}{120} f' p (3 q'^2) + \tfrac{2}{120} f^\circ (2 q^2) - \tfrac{2}{120} f' p (4 q'^2) \\ & - \tfrac{2}{120} f^\circ (3 q q') - \tfrac{2}{120} f^\circ (6 q q') - \tfrac{6}{120} g^\circ q (3 p^2) - \tfrac{2}{120} f^\circ (q q') - \tfrac{6}{120} g^\circ q' (3 p^2) \\ & - \tfrac{2}{120} f^\circ (3 q'^2) - \tfrac{2}{120} f^\circ (3 q'^2) + \tfrac{6}{120} g^\circ q (6 q^2) - \tfrac{2}{120} f^\circ (4 q'^2) + \tfrac{6}{120} g^\circ q' (2 q^2) \\ & \qquad - \tfrac{2}{120} f' p (3 p^2) - \tfrac{6}{120} g^\circ q (6 q q') - \tfrac{2}{120} f' p (3 p^2) - \tfrac{6}{120} g^\circ q' (q q') \\ & \qquad + \tfrac{2}{120} f' p (6 q^2) - \tfrac{6}{120} g^\circ q (3 q'^2) + \tfrac{2}{120} f' p (2 q^2) - \tfrac{6}{120} g^\circ q' (4 q'^2)\big). \end{aligned}$$

We know, further, that

$$k = -\frac{1}{n} \cdot \frac{\partial^2 n}{\partial q^2} = -2 f - 6 g q - (12 h - 2 f^2) q^2 - \text{etc.},$$

$$f = f^\circ + f' p + f'' p^2 + \text{etc.},$$
$$g = g^\circ + g' p + g'' p^2 + \text{etc.},$$
$$h = h^\circ + h' p + h'' p^2 + \text{etc.}$$

Hence, substituting these values for f, g, and h in k, we have at B where $k = \beta$, correct to terms of the third degree,

$$\beta = -2f^\circ - 2f' p - 6g^\circ q - 2f'' p^2 - 6g' pq - (12h^\circ - 2f^{\circ 2}) q^2.$$

Likewise, remembering that q becomes q' at C, and that both p and q vanish at A, we have

$$\gamma = -2f^\circ - 2f' p - 6g^\circ q' - 2f'' p^2 - 6g' pq' - (12h^\circ - 2f^{\circ 2}) q'^2,$$
$$\alpha = -2f^\circ.$$

And since $c \sin B = r \sin \psi$,

$$c \sin B = p(1 - \tfrac{1}{3}f^\circ q^2 - \tfrac{1}{4}f' pq^2 - \tfrac{1}{2}g^\circ q^3 - \text{etc.}).$$

Now, if we substitute in (1) $c \sin B$, α, β, γ for the series which they represent, and a for $q - q'$, we obtain (still correct to terms of the sixth degree)

$$\begin{aligned}\sigma = \tfrac{1}{2} ac \sin B \,\big(1 &+ \tfrac{1}{120}\alpha(4p^2 - 2q^2 + 3qq' + 3q'^2)\\ &+ \tfrac{1}{120}\beta(3p^2 - 6q^2 + 6qq' + 3q'^2)\\ &+ \tfrac{1}{120}\gamma(3p^2 - 2q^2 + qq' + 4q'^2)\big).\end{aligned}$$

And if in this equation we replace p, q, q' by $c \sin B$, $c \cos B$, $c \cos B - a$, respectively, we shall have

$$[8] \qquad \begin{aligned}\sigma = \tfrac{1}{2} ac \sin B \,\big(1 &+ \tfrac{1}{120}\alpha(3a^2 + 4c^2 - 9ac \cos B)\\ &+ \tfrac{1}{120}\beta(3a^2 + 3c^2 - 12ac \cos B)\\ &+ \tfrac{1}{120}\gamma(4a^2 + 3c^2 - 9ac \cos B)\big).\end{aligned}$$

By writing for B, α, β, a in [8], A, β, α, b respectively, we obtain at once formula [9]. Likewise by writing for B, β, γ, c in [8], C, γ, β, b respectively, we obtain formula [10]. Formulæ [9] and [10] can, of course, also be derived by the method used to derive [8].

34. Art. 26, p. 41, l. 13. The right hand side of this equation should have the positive sign. All the editions prior to Wangerin's have the incorrect sign.

35, Art. 26, p. 41. Derivation of formula [11].

We have

$$(1) \qquad \begin{aligned} &r^2 + r'^2 - (q - q')^2 - 2r \cos\phi \,.\, r' \cos\phi' - 2r \sin\phi \,.\, r' \sin\phi'\\ &= b^2 + c^2 - a^2 - 2bc \cos(\phi - \phi')\\ &= 2bc(\cos A^* - \cos A),\end{aligned}$$

since $b^2 + c^2 - a^2 = 2bc \cos A^*$ and $\cos(\phi - \phi') = \cos A$.

By priming the q's in formulæ [1], [4], [5] we obtain at once series for r'^2, $r'\cos\phi'$, $r'\sin\phi'$. Hence we have series for all the terms in the above expression, and also for the terms in the expression:

$$(2)\qquad r\sin\phi \,.\, r'\cos\phi' - r\cos\phi \,.\, r'\sin\phi' = bc\sin A,$$

namely,

$$\begin{aligned}(3)\qquad r^2 = p^2 &+ \tfrac{2}{3}f^\circ p^2 q^2 + \tfrac{1}{2}f' p^3 q^2 + (\tfrac{2}{5}f'' - \tfrac{4}{45}f^{\circ 2})p^4 q^2 + \text{etc.}\\ + q^2 &\qquad\qquad + \tfrac{1}{2}g^\circ p^2 q^3 + \tfrac{2}{5}g' p^3 q^3\\ &\qquad\qquad + (\tfrac{2}{5}h^\circ - \tfrac{7}{45}f^{\circ 2})p^2 q^4,\end{aligned}$$

$$\begin{aligned}(4)\qquad r'^2 = p^2 &+ \tfrac{2}{3}f^\circ p^2 q'^2 + \tfrac{1}{2}f' p^3 q'^2 + (\tfrac{2}{5}f'' - \tfrac{4}{45}f^{\circ 2})p^4 q'^2 + \text{etc.}\\ + q'^2 &\qquad\qquad + \tfrac{1}{2}g^\circ p^2 q'^3 + \tfrac{2}{5}g' p^3 q'^3\\ &\qquad\qquad + (\tfrac{2}{5}h^\circ - \tfrac{7}{45}f^{\circ 2})p^2 q'^4,\end{aligned}$$

$$(5)\qquad -(q-q')^2 = -q^2 + 2qq' - q'^2,$$

$$\begin{aligned}(6)\qquad 2r\cos\phi = 2p &+ \tfrac{4}{3}f^\circ p q^2 + \tfrac{10}{12}f' p^2 q^2 + (\tfrac{6}{10}f'' - \tfrac{16}{45}f^{\circ 2})p^3 q^2 + \text{etc.}\\ &+ g^\circ p q^3 \quad + \tfrac{14}{20}g' p^2 q^3\\ &\qquad\qquad + (\tfrac{4}{5}h^\circ - \tfrac{14}{45}f^{\circ 2})p q^4,\end{aligned}$$

$$\begin{aligned}(7)\qquad r'\cos\phi' = p &+ \tfrac{2}{3}f^\circ p q'^2 + \tfrac{5}{12}f' p^2 q'^2 + (\tfrac{3}{10}f'' - \tfrac{8}{45}f^{\circ 2})p^3 q'^2 + \text{etc.}\\ &+ \tfrac{1}{2}g^\circ p q'^3 + \tfrac{7}{20}g' p^2 q'^3\\ &\qquad\qquad + (\tfrac{2}{5}h^\circ - \tfrac{7}{45}f^{\circ 2})p q'^4,\end{aligned}$$

$$\begin{aligned}(8)\qquad 2r\sin\phi = 2q &- \tfrac{2}{3}f^\circ p^2 q - \tfrac{2}{6}f' p^3 q - (\tfrac{2}{10}f'' - \tfrac{14}{90}f^{\circ 2})p^4 q - \text{etc.}\\ &- \tfrac{2}{4}g^\circ p^2 q^2 - \tfrac{6}{20}g' p^3 q^2\\ &\qquad\qquad - (\tfrac{2}{5}h^\circ + \tfrac{26}{90}f^{\circ 2})p^2 q^3,\end{aligned}$$

$$\begin{aligned}(9)\qquad r'\sin\phi' = q' &- \tfrac{1}{3}f^\circ p^2 q' - \tfrac{1}{6}f' p^3 q' - (\tfrac{1}{10}f'' - \tfrac{7}{90}f^{\circ 2})p^4 q' - \text{etc.}\\ &- \tfrac{1}{4}g^\circ p^2 q'^2 - \tfrac{3}{20}g' p^3 q'^2\\ &\qquad\qquad - (\tfrac{1}{5}h^\circ + \tfrac{13}{90}f^{\circ 2})p^2 q'^3.\end{aligned}$$

By adding (3), (4), and (5), we obtain

$$\begin{aligned}(10)\quad r^2 + r'^2 - (q-q')^2 = 2p^2 &+ \tfrac{2}{3}f^\circ p^2(q^2+q'^2) + \tfrac{1}{2}f' p^3(q^2+q'^2) + (\tfrac{2}{5}f'' - \tfrac{4}{45}f^{\circ 2})p^4(q^2+q'^2) + \text{etc.}\\ + 2qq' &\qquad + \tfrac{1}{2}g^\circ p^2(q^3+q'^3) + \tfrac{2}{5}g' p^3(q^3+q'^3)\\ &\qquad\qquad + (\tfrac{2}{5}h^\circ - \tfrac{7}{45}f^{\circ 2})p^2(q^4+q'^4).\end{aligned}$$

On multiplying (6) by (7), we obtain

$$\begin{aligned}(11)\qquad &2r\cos\phi \,.\, r'\cos\phi'\\ &= 2p^2 + \tfrac{4}{3}f^\circ p^2(q^2+q'^2) + \tfrac{5}{6}f' p^3(q^2+q'^2) + (\tfrac{3}{5}f'' - \tfrac{16}{45}f^{\circ 2})p^4(q^2+q'^2) + \text{etc.}\\ &\qquad + g^\circ p^2(q^3+q'^3) + \tfrac{7}{10}g' p^3(q^3+q'^3)\\ &\qquad\qquad + (\tfrac{4}{5}h^\circ - \tfrac{14}{45}f^{\circ 2})p^2(q^4+q'^4)\\ &\qquad\qquad + \tfrac{8}{9}f^{\circ 2}p^2 q^2 q'^2,\end{aligned}$$

and multiplying (8) by (9), we obtain

$$
\begin{aligned}
(12)\qquad & 2\, r \sin\phi \,.\, r' \sin\phi' \\
&= 2\, q q' - \tfrac{4}{3} f^{\circ} p^2 q q' - \tfrac{2}{3} f' p^3 q q' - \left(\tfrac{2}{5} f'' - \tfrac{24}{45} f^{\circ 2}\right) p^4 q q' - \text{etc.} \\
&\qquad - \tfrac{1}{2} g^{\circ} p^2 q q' (q + q') - \tfrac{3}{10} g' p^3 q q' (q + q') \\
&\qquad - \left(\tfrac{2}{5} h^{\circ} + \tfrac{13}{45} f^{\circ 2}\right) p^2 q q' (q^2 + q'^2).
\end{aligned}
$$

Hence by adding (11) and (12), we have

$$
\begin{aligned}
(13)\qquad & 2\, b c \cos A \\
&= 2\, p^2 + \tfrac{4}{3} f^{\circ} p^2 (q^2 + q'^2) + \tfrac{1}{6} f' p^3 (5\, q^2 - 4\, q q' + 5\, q'^2) - \tfrac{8}{45} f^{\circ 2} p^4 (2\, q^2 + 2\, q'^2 - 3\, q q') + \text{etc.} \\
&\quad + 2 q q' - \tfrac{4}{3} f^{\circ} p^2 q q' + \tfrac{1}{2} g^{\circ} p^2 (2\, q^3 + 2\, q'^3 - q^2 q' - q q'^2) \\
&\quad - \tfrac{1}{45} f^{\circ 2} p^2 (14\, q^4 + 14\, q'^4 + 13 q^3 q' + 13\, q q'^3 - 40\, q^2 q'^2) \\
&\quad + \tfrac{1}{10} g' p^3 (7\, q^3 + 7\, q'^3 - 3\, q^2 q' - 3\, q q'^2) \\
&\quad + \tfrac{1}{5} f'' p^4 (3\, q^2 + 3\, q'^2 - 2\, q q') \\
&\quad + \tfrac{2}{5} h^{\circ} p^2 (2\, q^4 + 2\, q'^4 - q^3 q' - q q'^3).
\end{aligned}
$$

Therefore we have, by subtracting (13) from (10),

$$
\begin{aligned}
& 2\, b c\, (\cos A^* - \cos A) \\
&= - \tfrac{2}{3} f^{\circ} p^2 (q^2 + q'^2 - 2 q q') - \tfrac{1}{3} f' p^3 (q^2 + q'^2 - 2\, q q') + \tfrac{4}{15} f^{\circ 2} p^4 (q^2 + q'^2 - 2\, q q') - \text{etc.} \\
&\quad - \tfrac{1}{2} g^{\circ} p^2 (q^3 + q'^3 - q^2 q' - q q'^2) - \tfrac{1}{5} f'' p^4 (q^2 + q'^2 - 2\, q q') \\
&\quad + \tfrac{1}{45} f^{\circ 2} p^2 (7\, q^4 + 7\, q'^4 + 13\, q^3 q' + 13\, q q'^3 - 40\, q^2 q'^2) \\
&\quad - \tfrac{2}{5} h^{\circ} p^2 (q^4 + q'^4 - q^3 q' - q q'^3) \\
&\quad - \tfrac{3}{10} g' p^3 (q^3 + q'^3 - q^2 q' - q q'^2),
\end{aligned}
$$

which we can write thus:

$$
\begin{aligned}
(14)\quad 2\, b c\, (\cos A^* - \cos A) = -2\, p^2 (q - q')^2 \Big(& \tfrac{1}{3} f^{\circ} + \tfrac{1}{6} f' p + \tfrac{1}{4} g^{\circ} (q + q') + \tfrac{1}{10} f'' p^2 \\
& + \tfrac{1}{5} h^{\circ} (q^2 + q q' + q'^2) + \tfrac{3}{20} g' p\, (q + q') \\
& - \tfrac{2}{15} f^{\circ 2} p^2 - \tfrac{1}{90} f^{\circ 2} (7\, q^2 + 7\, q'^2 + 27\, q q') \Big),
\end{aligned}
$$

correct to terms of the seventh degree.

If we multiply (7) by [5] on page 37, we obtain

$$
\begin{aligned}
(15)\qquad & r \sin\phi \,.\, r' \cos\phi' \\
&= p q + \tfrac{2}{3} f^{\circ} p q q'^2 + \tfrac{5}{12} f' p^2 q q'^2 + \left(\tfrac{3}{10} f'' - \tfrac{8}{45} f^{\circ 2}\right) p^3 q q'^2 - \text{etc.} \\
&\quad - \tfrac{1}{3} f^{\circ} p^3 q + \tfrac{1}{2} g^{\circ} p q q'^3 + \tfrac{7}{20} g' p^2 q q'^3 \\
&\quad - \tfrac{1}{6} f' p^4 q + \left(\tfrac{2}{5} h^{\circ} - \tfrac{7}{45} f^{\circ 2}\right) p q q'^4 \\
&\quad - \tfrac{1}{4} g^{\circ} p^3 q^2 - \tfrac{2}{9} f^{\circ 2} p^3 q q'^2 \\
&\quad - \left(\tfrac{1}{10} f'' - \tfrac{7}{90} f^{\circ 2}\right) p^5 q \\
&\quad - \tfrac{3}{20} g' p^4 q^2 \\
&\quad - \left(\tfrac{1}{5} h^{\circ} + \tfrac{13}{90} f^{\circ 2}\right) p^3 q^3.
\end{aligned}
$$

And multiplying (9) by formula [4] on page 37, we obtain

$$(16)\qquad \begin{aligned} &r\cos\phi\,.\,r'\sin\phi' \\ &= p\,q' - \tfrac{1}{3}f^\circ p^3 q' - \tfrac{1}{6}f'p^4 q' - (\tfrac{1}{10}f'' - \tfrac{7}{90}f^{\circ 2})\,p^5 q' + \text{etc.} \\ &\qquad + \tfrac{2}{3}f^\circ p\,q^2 q' - \tfrac{1}{4}g^\circ p^3 q'^2 - \tfrac{3}{20}g' p^4 q'^2 \\ &\qquad\qquad + \tfrac{5}{12}f' p^2 q^2 q' - (\tfrac{1}{5}h^\circ + \tfrac{13}{90}f^{\circ 2})\,p^3 q'^3 \\ &\qquad\qquad + \tfrac{1}{2}g^\circ p\,q^3 q' - \tfrac{2}{9}f^{\circ 2} p^3 q^2 q' \\ &\qquad\qquad\qquad + (\tfrac{3}{10}f'' - \tfrac{8}{45}f^{\circ 2})\,p^3 q^2 q' \\ &\qquad\qquad\qquad + \tfrac{7}{20}g' p^2 q^3 q' \\ &\qquad\qquad\qquad + (\tfrac{2}{5}h^\circ - \tfrac{7}{45}f^{\circ 2})\,p\,q^4 q'. \end{aligned}$$

Therefore we have, by subtracting (16) from (15),

$$(17)\qquad \begin{aligned} &b\,c\sin A \\ &= p\,(q - q')\Big(1 - \tfrac{1}{3}f^\circ p^2 - \tfrac{5}{12}f' p\,q\,q' - (\tfrac{3}{10}f'' - \tfrac{8}{45}f^{\circ 2})\,p^2 q\,q' \\ &\qquad - \tfrac{2}{3}f^\circ q\,q' - \tfrac{1}{6}f' p^3 - (\tfrac{1}{10}f'' - \tfrac{7}{90}f^{\circ 2})\,p^4 \\ &\qquad\qquad - \tfrac{1}{2}g^\circ q\,q'\,(q + q') - \tfrac{7}{20}g' p\,q\,q'\,(q + q') \\ &\qquad\qquad - \tfrac{1}{4}g^\circ p^2\,(q + q') - \tfrac{3}{20}g' p^3\,(q + q') \\ &\qquad\qquad\qquad - (\tfrac{1}{5}h^\circ + \tfrac{13}{90}f^{\circ 2})\,p^2\,(q^2 + q\,q' + q'^2) \\ &\qquad\qquad\qquad - (\tfrac{2}{5}h^\circ - \tfrac{7}{45}f^{\circ 2})\,q\,q'\,(q^2 + q\,q' + q'^2) \\ &\qquad\qquad\qquad + \tfrac{2}{9}f^{\circ 2} p^2 q\,q'\Big), \end{aligned}$$

correct to terms of the seventh degree.

Let $A^* - A = \zeta$, whence $A^* = A + \zeta$, ζ being a magnitude of the second order. Hence we have, expanding $\sin\zeta$ and $\cos\zeta$, and rejecting powers of ζ above the second,

$$\cos A^* = \cos A\,.\left(1 - \frac{\zeta^2}{2}\right) - \sin A\,.\,\zeta,$$

or

$$\cos A^* - \cos A = -\frac{\cos A}{2}\cdot\zeta^2 - \sin A\,.\,\zeta\,;$$

or, multiplying both members of this equation by $2\,b\,c$,

$$(18)\qquad 2\,b\,c\,(\cos A^* - \cos A) = -\,b\,c\cos A\,.\,\zeta^2 - 2\,b\,c\sin A\,.\,\zeta.$$

Further, let $\zeta = R_2 + R_3 + R_4 + \text{etc.}$, where the R's have the same meaning as before. If now we substitute in (18) for its various terms the series derived above, we shall have, on rejecting terms above the sixth degree,

$$\begin{aligned} &(p^2 + q\,q')\,R_2^2 + 2\,p\,(q - q')\big(1 - \tfrac{1}{3}f^\circ\,(p^2 + 2\,q\,q')\big)\,(R_2 + R_3 + R_4) \\ &= 2\,p^2\,(q - q')^2\big(\tfrac{1}{3}f^\circ + \tfrac{1}{6}f' p + \tfrac{1}{4}g^\circ\,(q + q') + \tfrac{1}{10}f'' p^2 + \tfrac{3}{20}g' p\,(q + q') \\ &\qquad + \tfrac{1}{5}h^\circ\,(q^2 + q\,q' + q'^2) - \tfrac{1}{90}f^{\circ 2}\,(12\,p^2 + 7\,q^2 + 7\,q'^2 + 27\,q\,q')\big). \end{aligned}$$

Equating terms of like powers, and solving for R_2, R_3, R_4, we find

$$R_2 = p\,(q-q')\,.\,\tfrac{1}{3}f^\circ, \quad R_3 = p\,(q-q')\left(\tfrac{1}{6}f'p + \tfrac{1}{4}g^\circ\,(q+q')\right),$$
$$R_4 = p\,(q-q')\left(\tfrac{1}{10}f''p^2 + \tfrac{3}{20}g'p\,(q+q') + \tfrac{1}{5}h^\circ\,(q^2+qq'+q'^2) - \tfrac{1}{90}f^{\circ 2}\,(7p^2+7q^2+7q'^2+12qq')\right).$$

Therefore we have

$$\begin{aligned} A^* - A = p\,(q-q')\Big(&\tfrac{1}{3}f^\circ + \tfrac{1}{6}f'p + \tfrac{1}{4}g^\circ\,(q+q') + \tfrac{1}{10}f''p^2 \\ &+ \tfrac{3}{20}g'p\,(q+q') + \tfrac{1}{5}h^\circ\,(q^2+qq'+q'^2) \\ &- \tfrac{1}{90}f^{\circ 2}\,(7p^2+7q^2+12qq'+7q'^2)\Big), \end{aligned}$$

correct terms of the fifth degree.

This equation may be written as follows:

$$\begin{aligned} A^* = A + ap\left(1 - \tfrac{1}{6}f^\circ\,(p^2+q^2+q'^2+qq')\right)\Big(&\tfrac{1}{3}f^\circ + \tfrac{1}{6}f'p + \tfrac{1}{4}g^\circ\,(q+q') \\ &+ \tfrac{1}{10}f''p^2 + \tfrac{3}{20}g'p\,(q+q') + \tfrac{1}{5}h^\circ\,(q^2+qq'+q'^2) - \tfrac{1}{90}f^{\circ 2}(2p^2+2q^2+7qq'+2q'^2)\Big). \end{aligned}$$

But, since

$$2\,\sigma = ap\left(1 - \tfrac{1}{6}f^\circ\,(p^2+q^2+qq'+q'^2) + \text{etc.}\right),$$

the above equation becomes

$$\begin{aligned} A^* = A - \sigma\Big(&-\tfrac{2}{3}f^\circ - \tfrac{1}{3}f'p - \tfrac{1}{2}g^\circ\,(q+q') - \tfrac{1}{5}f''p^2 - \tfrac{3}{10}g'p\,(q+q') \\ &- \tfrac{2}{5}h^\circ\,(q^2+qq'+q'^2) + \tfrac{1}{90}f^{\circ 2}\,(4p^2+4q^2+14qq'+4q'^2)\Big), \end{aligned}$$

or

$$\begin{aligned} A^* = A - \sigma\Big(-\tfrac{2}{6}f^\circ &- \tfrac{2}{12}f^\circ \quad - \tfrac{2}{12}f^\circ \\ &- \tfrac{2}{12}f'p \quad - \tfrac{2}{12}f'p \\ &- \tfrac{6}{12}g^\circ q \quad - \tfrac{6}{12}g^\circ q' \\ &- \tfrac{2}{12}f''p^2 - \tfrac{2}{12}f''p^2 + \tfrac{2}{15}f''p^2 \\ &- \tfrac{6}{12}g'pq - \tfrac{6}{12}g'pq' + \tfrac{1}{5}g'p\,(q+q') \\ &- \tfrac{12}{12}h^\circ q^2 - \tfrac{12}{12}h^\circ q'^2 + \tfrac{1}{5}h^\circ\,(3q^2-2qq'+3q'^2) \\ &+ \tfrac{2}{12}f^{\circ 2}q^2 + \tfrac{2}{12}f^{\circ 2}q'^2 + \tfrac{1}{90}f^{\circ 2}\,(4p^2-11q^2+14qq'-11q'^2)\Big). \end{aligned}$$

Therefore, if we substitute in this equation α, β, γ for the series which they represent, we shall have

$$\begin{aligned} \text{[11]} \qquad A^* = A - \sigma\Big(&\tfrac{1}{6}\alpha + \tfrac{1}{12}\beta + \tfrac{1}{12}\gamma + \tfrac{2}{15}f''p^2 + \tfrac{1}{5}g'p\,(q+q') \\ &+ \tfrac{1}{5}h^\circ\,(3q^2-2qq'+3q'^2) + \tfrac{1}{90}f^{\circ 2}\,(4p^2-11q^2+14qq'-11q'^2)\Big). \end{aligned}$$

36. Art. 26, p. 41. Derivation of formula [12].

We form the expressions $(q-q')^2 + r^2 - r'^2 - 2\,(q-q')\,r\cos\psi$ and $(q-q')\,r\sin\psi$. Then, since

$$(q-q')^2 + r^2 - r'^2 = a^2 + c^2 - b^2 = 2\,ac\cos B^*,$$
$$2\,(q-q')\,r\cos\psi = 2\,ac\cos B,$$

we have

$$(q-q')^2+r^2-r'^2-2(q-q')\,r\cos\psi=2\,a\,c\,(\cos B^*-\cos B).$$

We have also

$$(q-q')\,r\sin\psi=ac\sin B.$$

Subtracting (4) on page 68 from [1] on page 36, and adding this difference to $(q-q')^2$, we obtain

(1) $(q-q')^2+r^2-r'^2$, or $2\,ac\cos B^*$

$$\begin{aligned}=2\,q(q-q')+\tfrac{2}{3}f^\circ p^2(q^2-q'^2)+\tfrac{1}{2}f'p^3(q^2-q'^2)&+(\tfrac{2}{5}f''-\tfrac{4}{45}f^{\circ 2})p^4(q^2-q'^2)+\text{etc.}\\ +\tfrac{1}{2}g^\circ p^2(q^3-q'^3)+\tfrac{2}{5}g'p^3(q^3-q'^3)&\\ &+(\tfrac{2}{5}h^\circ-\tfrac{7}{45}f^{\circ 2})p^2(q^4-q'^4).\end{aligned}$$

If we multiply [3] on page 37 by $2(q-q')$, we obtain

(2) $2(q-q')\,r\cos\psi$, or $2\,ac\cos B$

$$\begin{aligned}=2\,q\,(q-q')+\tfrac{4}{3}f^\circ p^2q(q-q')+f'p^3q(q-q')&+(\tfrac{4}{5}f''-\tfrac{8}{45}f^{\circ 2})p^4q\,(q-q')+\text{etc.}\\ +\tfrac{3}{2}g^\circ p^2q^2(q-q')+\tfrac{6}{5}g'p^3q^2(q-q')&\\ &+(\tfrac{8}{5}h^\circ-\tfrac{28}{45}f^{\circ 2})p^2q^3(q-q').\end{aligned}$$

Subtracting (2) from (1), we have

(3) $2\,ac\,(\cos B^*-\cos B)$

$$\begin{aligned}=-2\,p^2(q-q')^2\Big(\tfrac{1}{3}f^\circ+\tfrac{1}{4}f'p&+(\tfrac{1}{5}f''-\tfrac{2}{45}f^{\circ 2})p^2+\text{etc.}\\ +\tfrac{1}{4}g^\circ(2\,q+q')+\tfrac{1}{5}g'p(2\,q+q')&\\ &+(\tfrac{1}{5}h^\circ-\tfrac{7}{90}f^{\circ 2})(3\,q^2+2\,qq'+q'^2)\Big).\end{aligned}$$

Multiplying [2] on page 36 by $(q-q')$, we obtain at once

(4) $(q-q')\,r\sin\psi$, or $ac\sin B$

$$\begin{aligned}=p(q-q')\Big(1-\tfrac{1}{3}f^\circ q^2-\tfrac{1}{4}f'pq^2-(\tfrac{1}{5}f''+\tfrac{8}{45}f^{\circ 2})p^2q^2+\text{etc.}\\ -\tfrac{1}{2}g^\circ q^3-\tfrac{2}{5}g'pq^3\\ -(\tfrac{2}{5}h^\circ-\tfrac{8}{45}f^{\circ 2})q^4\Big).\end{aligned}$$

We now set $B^*-B=\zeta$, whence $B^*=B+\zeta$, and therefore

$$\cos B^*=\cos B\cos\zeta-\sin B\sin\zeta.$$

This becomes, after expanding $\cos\zeta$ and $\sin\zeta$ and neglecting powers of ζ above the second,

$$\cos B^*-\cos B=-\frac{\cos B}{2}\cdot\zeta^2-\sin B\,.\,\zeta.$$

Multiplying both members of this equation by $2\,ac$, we obtain

(5) $$2\,ac(\cos B^*-\cos B)=-ac\cos B\,.\,\zeta^2-2\,ac\sin B\,.\,\zeta.$$

Again, let $\zeta = R_2 + R_3 + R_4 +$ etc., where the R's have the same meaning as before. Hence, replacing the terms in (5) by the proper series and neglecting terms above the sixth degree, we have

$$(6)\qquad \begin{aligned} &q(q-q')R_2^2 + 2p(q-q')(1-\tfrac{1}{3}f^\circ q^2)(R_2+R_3+R_4) \\ &= 2p^2(q-q')^2\big(\tfrac{1}{3}f^\circ + \tfrac{1}{4}f'p + (\tfrac{1}{5}f'' - \tfrac{2}{45}f^{\circ 2})p^2 \\ &\quad + \tfrac{1}{4}g^\circ(2q+q') + \tfrac{1}{5}g'p(2q+q') \\ &\quad + (\tfrac{1}{5}h^\circ - \tfrac{7}{90}f^{\circ 2})(3q^2+2qq'+q'^2)\big). \end{aligned}$$

From this equation we find

$$\begin{aligned} &R_2 = p(q-q')\cdot\tfrac{1}{3}f^\circ, \quad R_3 = p(q-q')\big(\tfrac{1}{4}f'p + \tfrac{1}{4}g^\circ(2q+q')\big), \\ &R_4 = p(q-q')\big(\tfrac{1}{5}f''p^2 + \tfrac{1}{5}g'p(2q+q') + \tfrac{1}{5}h^\circ(3q^2+2qq'+q'^2) \\ &\qquad - \tfrac{1}{90}f^{\circ 2}(4p^2+16q^2+9qq'+7q'^2)\big). \end{aligned}$$

Therefore we have, correct to terms of the fifth degree,

$$\begin{aligned} B^* - B = p(q-q')\big(&\tfrac{1}{3}f^\circ + \tfrac{1}{4}f'p + \tfrac{1}{5}f''p^2 + \tfrac{1}{5}g'p(2q+q') \\ &+ \tfrac{1}{4}g^\circ(2q+q') + \tfrac{1}{5}h^\circ(3q^2+2qq'+q'^2) \\ &- \tfrac{1}{90}f^{\circ 2}(4p^2+16q^2+9qq'+7q'^2)\big), \end{aligned}$$

or, after factoring the last factor on the right,

$$(7)\qquad \begin{aligned} B^* = B - \tfrac{1}{2}p(q-q')\big(1-\tfrac{1}{6}f^\circ(p^2+q^2+qq'+q'^2)\big)\big(&-\tfrac{2}{3}f^\circ - \tfrac{1}{2}f'p - \tfrac{1}{2}g^\circ(2q+q') \\ &- \tfrac{2}{5}f''p^2 - \tfrac{2}{5}g'p(2q+q') - \tfrac{2}{5}h^\circ(3q^2+2qq'+q'^2) \\ &+ \tfrac{1}{90}f^{\circ 2}(-2p^2+22q^2+8qq'+4q'^2)\big). \end{aligned}$$

The last factor on the right in (7) may be put in the form:

$$\begin{aligned} \big(-\tfrac{2}{12}f^\circ &- \tfrac{2}{6}f^\circ - \tfrac{2}{12}f^\circ \\ &- \tfrac{2}{6}f'p - \tfrac{2}{12}f'p \\ &- \tfrac{6}{6}g^\circ q - \tfrac{6}{12}g^\circ q' \\ &- \tfrac{2}{6}f''p^2 - \tfrac{2}{12}f''p^2 + \tfrac{1}{10}f''p^2 \\ &- \tfrac{6}{6}g'pq - \tfrac{6}{12}g'pq' + \tfrac{1}{10}g'p(2q+q') \\ &- \tfrac{12}{6}h^\circ q^2 - \tfrac{12}{12}h^\circ q'^2 + \tfrac{1}{5}h^\circ(4q^2+3q'^2-4qq') \\ &+ \tfrac{2}{6}f^{\circ 2}q^2 + \tfrac{2}{12}f^{\circ 2}q'^2 - \tfrac{1}{90}f^{\circ 2}(2p^2+8q^2+11q'^2-8qq')\big). \end{aligned}$$

Finally, substituting in (7) σ, α, β, γ for the expressions which they represent, we obtain, still correct to terms of the fifth degree,

$$[12]\qquad \begin{aligned} B^* = B - \sigma\big(&\tfrac{1}{12}\alpha + \tfrac{1}{6}\beta + \tfrac{1}{12}\gamma + \tfrac{1}{10}f''p^2 \\ &+ \tfrac{1}{10}g'p(2q+q') + \tfrac{1}{5}h^\circ(4q^2-4qq'+3q'^2) \\ &- \tfrac{1}{90}f^{\circ 2}(2p^2+8q^2-8qq'+11q'^2)\big). \end{aligned}$$

37. Art. 26, p. 41. Derivation of formula [13].

Here we form the expressions $(q-q')^2+r'^2-r^2-2(q-q')\,r'\cos(\pi-\psi')$ and $(q-q')\,r'\sin(\pi-\psi')$ and expand them into series. Since

$$(q-q')^2+r'^2-r^2=a^2+b^2-c^2=2\,ab\cos C^*,$$
$$2\,(q-q')\,r'\cos(\pi-\psi')=2\,ab\cos C,$$

we have

$$(q-q')^2+r'^2-r^2-2\,(q-q')\,r'\cos(\pi-\psi')=2\,ab\,(\cos C^*-\cos C).$$

We have also

$$(q-q')\,r'\sin(\pi-\psi')=ab\sin C.$$

Subtracting (3) on page 68 from (4) on the same page, and adding the result to $(q-q')^2$, we find

$$\begin{aligned}(1)\quad &(q-q')^2+r'^2-r^2,\ \text{or}\ 2\,ab\cos C^*\\ &=-2\,q'(q-q')-\tfrac{2}{3}f^\circ p^2(q^2-q'^2)-\tfrac{1}{2}f'p^3(q^2-q'^2)-(\tfrac{2}{5}f''-\tfrac{4}{45}f^{\circ 2})p^4(q^2-q'^2)-\text{etc.}\\ &\qquad -\tfrac{1}{2}g^\circ p^2(q^3-q'^3)-\tfrac{2}{5}g'p^3(q^3-q'^3)\\ &\qquad -(\tfrac{2}{5}h^\circ-\tfrac{7}{45}f^{\circ 2})p^2(q^4-q'^4).\end{aligned}$$

By priming the q's in formula [3] on page 37, we get a series for $r'\cos\psi'$, or for $-r'\cos(\pi-\psi')$. If we multiply this series for $-r'\cos(\pi-\psi')$ by $2(q-q')$, we find

$$\begin{aligned}(2)\quad &-2\,(q-q')\,r'\cos(\pi-\psi'),\ \text{or}\ -2\,ab\cos C\\ &=2\,(q-q')\big(q'+\tfrac{2}{3}f^\circ p^2q'+\tfrac{1}{2}f'p^3q'+(\tfrac{2}{5}f''-\tfrac{4}{45}f^{\circ 2})p^4q'+\text{etc.}\\ &\qquad +\tfrac{3}{4}g^\circ p^2q'^2+\tfrac{3}{5}g'p^3q'^2\\ &\qquad +(\tfrac{4}{5}h^\circ-\tfrac{14}{45}f^{\circ 2})p^2q'^3\big).\end{aligned}$$

And therefore, by adding (1) and (2), we obtain

$$\begin{aligned}(3)\quad &2\,ab\,(\cos C^*-\cos C)\\ &=-2\,p^2(q-q')^2\big(\tfrac{1}{3}f^\circ+\tfrac{1}{4}f'p+(\tfrac{1}{5}f''-\tfrac{2}{45}f^{\circ 2})p^2+\text{etc.}\\ &\qquad +\tfrac{1}{4}g^\circ(q+2\,q')+\tfrac{1}{5}g'p\,(q+2\,q')\\ &\qquad +(\tfrac{1}{5}h^\circ-\tfrac{7}{90}f^{\circ 2})(q^2+2\,qq'+3\,q'^2)\big).\end{aligned}$$

By priming the q's in [2] on page 36, we obtain a series for $r'\sin\psi'$, or for $r'\sin(\pi-\psi')$. Then, multiplying this series for $r'\sin(\pi-\psi')$ by $(q-q')$, we find

$$\begin{aligned}(4)\quad &(q-q')\,r'\sin(\pi-\psi'),\ \text{or}\ ab\sin C\\ &=p\,(q-q')\big(1-\tfrac{1}{3}f^\circ q'^2-\tfrac{1}{4}f'pq'^2-(\tfrac{1}{5}f''+\tfrac{8}{45}f^{\circ 2})p^2q'^2-\text{etc.}\\ &\qquad -\tfrac{1}{2}g^\circ q'^3-\tfrac{2}{5}g'pq'^3\\ &\qquad -(\tfrac{3}{5}h^\circ-\tfrac{8}{45}f^{\circ 2})\,q'^4\big).\end{aligned}$$

As before, let $C^*-C=\zeta$, whence $C^*=C+\zeta$, and therefore

$$\cos C^*=\cos C\cos\zeta-\sin C\sin\zeta.$$

Expanding $\cos\zeta$ and $\sin\zeta$ and neglecting powers of ζ above the second, this equation becomes

$$\cos C^* - \cos C = -\frac{\cos C}{2}\cdot\zeta^2 - \sin C\,.\,\zeta,$$

or, after multiplying both members by $2\,ab$,

$$2\,ab(\cos C^* - \cos C) = -ab\cos C\,.\,\zeta^2 - 2\,ab\sin C\,.\,\zeta. \tag{5}$$

Again we put $\zeta = R_2 + R_3 + R_4 +$ etc., the R's having the same meaning as before. Now, by substituting (2), (3), (4) in (5), and omitting terms above the sixth degree, we obtain

$$\begin{aligned} q'(q-q')R_2^2 - 2p(q-q')(1-\tfrac{1}{3}f^\circ q'^2)(R_2+R_3+R_4) \\ = -2p^2(q-q')^2\Big(\tfrac{1}{3}f^\circ + \tfrac{1}{4}f'p \qquad &+ (\tfrac{1}{5}f'' - \tfrac{2}{45}f^{\circ 2})p^2 \\ + \tfrac{1}{4}g^\circ(q+2q') &+ \tfrac{1}{5}g'p(q+2q') \\ &+ (\tfrac{1}{5}h^\circ - \tfrac{7}{90}f^{\circ 2})(q^2+2qq'+3q'^2)\Big), \end{aligned}$$

from which we find

$$\begin{aligned} R_2 &= p(q-q')\cdot\tfrac{1}{3}f^\circ, \quad R_3 = p(q-q')\big(\tfrac{1}{4}f'p + \tfrac{1}{4}g^\circ(q+2q')\big), \\ R_4 &= p(q-q')\big(\tfrac{1}{5}f''p^2 + \tfrac{1}{5}g'p(q+2q') + \tfrac{1}{5}h^\circ(q^2+2qq'+3q'^2) \\ &\qquad - \tfrac{1}{90}f^{\circ 2}(4p^2+7q^2+9qq'+16q'^2)\big). \end{aligned}$$

Therefore we have, correct to terms of the fifth degree,

$$\begin{aligned} C^* - C = p(q-q')\Big(\tfrac{1}{3}f^\circ + \tfrac{1}{4}f'p \qquad &+ \tfrac{1}{5}f''p^2 + \tfrac{1}{5}g'p(q+2q') \\ + \tfrac{1}{4}g^\circ(q+2q') &+ \tfrac{1}{5}h^\circ(q^2+2qq'+3q'^2) \\ &- \tfrac{1}{90}f^{\circ 2}(4p^2+7q^2+9qq'+16q'^2)\Big). \end{aligned} \tag{6}$$

The last factor on the right in (6) may be written as the product of two factors, one of which is $\tfrac{1}{2}\big(1-\tfrac{1}{6}f^\circ(p^2+q^2+qq'+q'^2)\big)$, and the other,

$$\begin{aligned} 2\Big(\tfrac{1}{3}f^\circ + \tfrac{1}{4}f'p + \tfrac{1}{4}g^\circ(q+2q') + \tfrac{1}{5}f''p^2 + \tfrac{1}{5}g'p(q+2q') \\ + \tfrac{1}{5}h^\circ(q^2+3q'^2+2qq') - \tfrac{1}{90}f^{\circ 2}(-p^2+2q^2+4qq'+11q'^2)\Big), \end{aligned}$$

or, in another form,

$$\begin{aligned} -\Big(-\tfrac{2}{12}f^\circ &- \tfrac{2}{12}f^\circ \quad - \tfrac{2}{6}f^\circ \\ &- \tfrac{2}{12}f'p \quad - \tfrac{2}{6}f'p \\ &- \tfrac{6}{12}g^\circ q \quad - \tfrac{6}{6}g^\circ q' \\ &- \tfrac{2}{12}f''p^2 - \tfrac{2}{6}f''p^2 \quad + \tfrac{1}{10}f''p^2 \\ &- \tfrac{6}{12}g'pq - \tfrac{6}{6}g'pq' \quad + \tfrac{1}{10}g'p(q+2q') \\ &- \tfrac{12}{12}h^\circ q^2 - \tfrac{12}{6}h^\circ q'^2 + \tfrac{1}{5}h^\circ(3q^2-4qq'+4q'^2) \\ &+ \tfrac{2}{12}f^{\circ 2}q^2 + \tfrac{2}{6}f^{\circ 2}q'^2 \quad - \tfrac{1}{90}f^{\circ 2}(2p^2+11q^2-8qq'+8q'^2)\Big). \end{aligned}$$

Hence (6) becomes, on substituting σ, α, β, γ for the expressions which they represent,

$$[13] \qquad \begin{aligned} C^* = C - \sigma\Big(&\tfrac{1}{12}\alpha + \tfrac{1}{12}\beta + \tfrac{1}{6}\gamma + \tfrac{1}{10}f''p^2 \\ &+ \tfrac{1}{10}g'p(q+2q') + \tfrac{1}{5}h^\circ(3q^2 - 4qq' + 4q'^2) \\ &- \tfrac{1}{90}f^{\circ 2}(2p^2 + 11q^2 - 8qq' + 8q'^2)\Big). \end{aligned}$$

38. Art. 26, p. 41. Derivation of formula [14].

This formula is derived at once by adding formulæ [11], [12], [13]. But, as Gauss suggests, it may also be derived from [6], p. 38. By priming the q's in [6] we obtain a series for $(\psi' + \phi')$. Subtracting this series from [6], and noting that $\phi - \phi' + \psi + \pi - \psi' = A + B + C$, we have, correct terms of the fifth degree,

$$(1) \qquad \begin{aligned} A+B+C = \pi - p(q-q')\Big(&f^\circ + \tfrac{2}{3}f'p + \tfrac{1}{2}f''p^2 + \tfrac{3}{4}g'p(q+q') \\ &+ g^\circ(q+q') + h^\circ(q^2 + qq' + q'^2) \\ &- \tfrac{1}{6}f^{\circ 2}(p^2 + 2q^2 + 2qq' + 2q'^2)\Big). \end{aligned}$$

The second term on the right in (1) may be written

$$\begin{aligned} +\tfrac{1}{2}ap\Big(1 - \tfrac{1}{6}f^\circ(p^2 + q^2 + qq' + q'^2)\Big)\cdot 2\Big(&-f^\circ - \tfrac{2}{3}f'p - \tfrac{1}{2}f''p^2 - \tfrac{3}{4}g'p(q+q') \\ &- g^\circ(q+q') - h^\circ(q^2 + qq' + q'^2) \\ &+ \tfrac{1}{6}f^{\circ 2}(+q^2 + qq' + q'^2)\Big), \end{aligned}$$

of which the last factor may be thrown into the form:

$$\begin{aligned} \Big(-\tfrac{2}{3}f^\circ &- \tfrac{2}{3}f^\circ && - \tfrac{2}{3}f^\circ \\ &- \tfrac{2}{3}f'p && - \tfrac{2}{3}f'p \\ &- \tfrac{6}{3}g^\circ q && - \tfrac{6}{3}g^\circ q' \\ &- \tfrac{2}{3}f''p^2 && - \tfrac{2}{3}f''p^2 + \tfrac{1}{3}f''p^2 \\ &- \tfrac{6}{3}g'pq && - \tfrac{6}{3}g'pq' + \tfrac{1}{2}g'p(q+q') \\ &- \tfrac{12}{3}h^\circ q^2 && - \tfrac{12}{3}h^\circ q'^2 + 2h^\circ(q^2 + q'^2 - qq') \\ &+ \tfrac{2}{3}f^{\circ 2}q^2 && + \tfrac{2}{3}f^{\circ 2}q'^2 - \tfrac{1}{3}f^{\circ 2}(q^2 + q'^2 - qq')\Big). \end{aligned}$$

Hence, by substituting σ, α, β, γ for the expressions they represent, (1) becomes

$$[14] \qquad \begin{aligned} A+B+C = \pi + \sigma\Big(&\tfrac{1}{3}\alpha + \tfrac{1}{3}\beta + \tfrac{1}{3}\gamma + \tfrac{1}{3}f''p^2 \\ &+ \tfrac{1}{2}g'p(q+q') + (2h^\circ - \tfrac{1}{3}f^{\circ 2})(q^2 - qq' + q'^2)\Big). \end{aligned}$$

Art. 27, p. 42. Omitting terms above the second degree, we have

$$a^2 = q^2 - 2qq' + q'^2, \qquad b^2 = p^2 + q'^2, \qquad c^2 = p^2 + q^2.$$

The expressions in the parentheses of the first set of formulæ for A^*, B^*, C^* in Art. 27 may be arranged in the following manner:

$$\begin{aligned} (2p^2 - q^2 + 4qq' - q'^2) &= \big((p^2 + q'^2) + (p^2 + q^2) - 2(q^2 - 2qq' + q'^2)\big), \\ (p^2 - 2q^2 + 2qq' + q'^2) &= \big(2(p^2 + q'^2) - (p^2 + q^2) - (q^2 - 2qq' + q'^2)\big), \\ (p^2 + q^2 + 2qq' - 2q'^2) &= \big(-(p^2 + q'^2) + 2(p^2 + q^2) - (q^2 - 2qq' + q'^2)\big). \end{aligned}$$

Now substituting a^2, b^2, c^2 for $(q^2 - 2\,qq' + q'^2)$, $(p^2 + q'^2)$, $(p^2 + q^2)$ respectively, and changing the signs of both members of the last two of these equations, we have

$$\begin{aligned}(2\,p^2 - q^2 \;\; + 4\,qq' - q'^2) &= (b^2 + c^2 - 2\,a^2),\\ -(p^2 \;\; - 2\,q^2 + 2\,qq' + q'^2) &= (a^2 + c^2 - 2\,b^2),\\ -(p^2 \;\; + q^2 \;\; + 2\,qq' - 2\,q'^2) &= (a^2 + b^2 - 2\,c^2).\end{aligned}$$

And replacing the expressions in the parentheses in the first set of formulæ for A^*, B^*, C^* by their equivalents, we get the second set.

Art. 27, p. 42. $f^\circ = -\dfrac{1}{2\,R^2}$, $f'' = 0$, etc., may be obtained directly, without the use of the general considerations of Arts. 25 and 26, in the following way. In the case of the sphere

$$ds^2 = \cos^2\left(\frac{q}{R}\right) \cdot dp^2 + dq^2,$$

hence

$$n = \cos\left(\frac{q}{R}\right) = 1 - \frac{q^2}{2\,R^2} + \frac{q^4}{24\,R^4} - \text{etc.},$$

i. e.,

$$f^\circ = -\frac{1}{2R^2}, \qquad h^\circ = \frac{1}{24R^4}, \qquad f' = g^\circ = f'' = g' = 0.$$

[Wangerin.]

39. Art. 27, p. 42, l. 16. This theorem of Legendre is found in the Mémoires (Histoire) de l'Academie Royale de Paris, 1787, p. 358, and also in his Trigonometry, Appendix, § V. He states it as follows in his Trigonometry:

The very slightly curved spherical triangle, whose angles are A, B, C and whose sides are a, b, c, always corresponds to a rectilinear triangle, whose sides a, b, c are of the same lengths, and whose opposite angles are $A - \frac{1}{3}\,e$, $B - \frac{1}{3}\,e$, $C - \frac{1}{3}\,e$, e being the excess of the sum of the angles in the given spherical triangle over two right angles.

40. Art. 28, p. 43, l. 7. The sides of this triangle are Hohehagen-Brocken, Inselberg-Hohehagen, Brocken-Inselberg, and their lengths are about 107, 85, 69 kilometers respectively, according to Wangerin.

41. Art. 29, p. 43. Derivation of the relation between σ and σ^*.

In Art. 28 we found the relation

$$A^* = A - \tfrac{1}{12}\,\sigma(2\,\alpha + \beta + \gamma).$$

Therefore

$$\sin A^* = \sin A \cos\left(\tfrac{1}{12}\,\sigma(2\,\alpha + \beta + \gamma)\right) - \cos A \sin\left(\tfrac{1}{12}\,\sigma(2\,\alpha + \beta + \gamma)\right),$$

which, after expanding $\cos\left(\tfrac{1}{12}\,\sigma(2\,\alpha + \beta + \gamma)\right)$ and $\sin\left(\tfrac{1}{12}\,\sigma(2\,\alpha + \beta + \gamma)\right)$ and rejecting powers of $\left(\tfrac{1}{12}\,\sigma(2\,\alpha + \beta + \gamma)\right)$ above the first, becomes

(1) $$\sin A^* = \sin A - \cos A \,.\, (\tfrac{1}{12}\sigma(2\alpha + \beta + \gamma)),$$

correct to terms of the fourth degree.

But, since σ and σ^* differ only by terms above the second degree, we may replace in (1) σ by the value of σ^*, $\frac{1}{2}bc\sin A^*$. We thus obtain, with equal exactness,

(2) $$\sin A = \sin A^*(1 + \tfrac{1}{24}bc\cos A \,.\, (2\alpha + \beta + \gamma)).$$

Substituting this value for $\sin A$ in [9], p. 40, we have, correct to terms of the sixth degree, the first formula for σ given in Art. 29. Since $2bc\cos A^*$, or $b^2 + c^2 - a^2$, differs from $2bc\cos A$ only by terms above the second degree, we may replace $2bc\cos A$ in this formula for σ by $b^2 + c^2 - a^2$. Also $\sigma^* = \frac{1}{2}bc\sin A^*$. Hence, if we make these substitutions in the first formula for σ, we obtain the second formula for σ with the same exactness. In the case of a sphere, where $\alpha = \beta = \gamma$, the second formula for σ reduces to the third.

When the surface is spherical, (2) becomes

$$\sin A = \sin A^*(1 + \frac{\alpha}{6}bc\cos A).$$

And replacing $2bc\cos A$ in this equation by $(b^2 + c^2 - a^2)$, we have

$$\sin A = \sin A^*(1 + \frac{\alpha}{12}(b^2 + c^2 - a^2)),$$

or

$$\frac{\sin A}{\sin A^*} = (1 + \frac{\alpha}{12}(b^2 + c^2 - a^2)).$$

And likewise we can find

$$\frac{\sin B}{\sin B^*} = (1 + \frac{\alpha}{12}(a^2 + c^2 - b^2)), \qquad \frac{\sin C}{\sin C^*} = (1 + \frac{\alpha}{12}(a^2 + b^2 - c^2)).$$

Multiplying together the last three equations and rejecting the terms containing α^2 and α^3, we have

$$1 + \frac{\alpha}{12}(a^2 + b^2 + c^2) = \frac{\sin A \,.\, \sin B \,.\, \sin C}{\sin A^* \,.\, \sin B^* \,.\, \sin C^*}.$$

Finally, taking the square root of both members of this equation, we have, with the same exactness,

$$\sigma = 1 + \frac{\alpha}{24}(a^2 + b^2 + c^2) = \sqrt{\left(\frac{\sin A \,.\, \sin B \,.\, \sin C}{\sin A^* \,.\, \sin B^* \,.\, \sin C^*}\right)}.$$

The method here used to derive the last formula from the next to the last formula of Art. 29 is taken from Wangerin.

NEW GENERAL INVESTIGATIONS

OF

CURVED SURFACES

NEW GENERAL INVESTIGATIONS

OF

CURVED SURFACES

(The notes referred to in the text are to be found on pages 111 - 114.)

Although the real purpose of this work is the deduction of new theorems concerning its subject, nevertheless we shall first develop what is already known, partly for the sake of consistency and completeness, and partly because our method of treatment is different from that which has been used heretofore. We shall even begin by advancing certain properties concerning plane curves from the same principles.

1.

In order to compare in a convenient manner the different directions of straight lines in a plane with each other, we imagine a circle with unit radius described in the plane about an arbitrary centre. The position of the radius of this circle, drawn parallel to a straight line given in advance, represents then the position of that line. And the angle which two straight lines make with each other is measured by the angle between the two radii representing them, or by the arc included between their extremities. Of course, where precise definition is necessary, it is specified at the outset, for every straight line, in what sense it is regarded as drawn. Without such a distinction the direction of a straight line would always correspond to two opposite radii.

2.

In the auxiliary circle we take an arbitrary radius as the first, or its terminal point in the circumference as the origin, and determine the positive sense of measuring the arcs from this point (whether from left to right or the contrary); in the opposite direction the arcs are regarded then as negative. Thus every direction of a straight line is expressed in degrees, etc., or also by a number which expresses them in parts of the radius.

Such lines as differ in direction by 360°, or by a multiple of 360°, have, therefore, precisely the same direction, and may, generally speaking, be regarded as the same. However, in such cases where the manner of describing a variable angle is taken into consideration, it may be necessary to distinguish carefully angles differing by 360°.

If, for example, we have decided to measure the arcs from left to right, and if to two straight lines l, l' correspond the two directions L, L', then $L'-L$ is the angle between those two straight lines. And it is easily seen that, since $L'-L$ falls between $-180°$ and $+180°$, the positive or negative value indicates at once that l' lies on the right or the left of l, as seen from the point of intersection. This will be determined generally by the sign of $\sin(L'-L)$.

If aa' is a part of a curved line, and if to the tangents at a, a' correspond respectively the directions α, α', by which letters shall be denoted also the corresponding points on the auxiliary circles, and if A, A' be their distances along the arc from the origin, then the magnitude of the arc $\alpha\alpha'$ or $A'-A$ is called the *amplitude* of aa'.

The comparison of the amplitude of the arc aa' with its length gives us the notion of curvature. Let l be any point on the arc aa', and let λ, Λ be the same with reference to it that α, A and α', A' are with reference to a and a'. If now $\alpha\lambda$ or $\Lambda - A$ be proportional to the part al of the arc, then we shall say that aa' is uniformly curved throughout its whole length, and we shall call

$$\frac{\Lambda - A}{al}$$

the measure of curvature, or simply the curvature. We easily see that this happens only when aa' is actually the arc of a circle, and that then, according to our definition, its curvature will be $\pm \frac{1}{r}$, if r denotes the radius. Since we always regard r as positive, the upper or the lower sign will hold according as the centre lies to the right or to the left of the arc aa' (a being regarded as the initial point, a' as the end point, and the directions on the auxiliary circle being measured from left to right). Changing one of these conditions changes the sign, changing two restores it again.

On the contrary, if $\Lambda - A$ be not proportional to al, then we call the arc non-uniformly curved and the quotient

$$\frac{\Lambda - A}{al}$$

may then be called its mean curvature. Curvature, on the contrary, always presupposes that the point is determined, and is defined as the mean curvature of an element at this point; it is therefore equal to

$$\frac{d\Lambda}{d\,al}.$$

We see, therefore, that arc, amplitude, and curvature sustain a similar relation to each other as time, motion, and velocity, or as volume, mass, and density. The reciprocal of the curvature, namely,

$$\frac{d\,al}{d\Lambda},$$

is called the radius of curvature at the point l. And, in keeping with the above conventions, the curve at this point is called concave toward the right and convex toward the left, if the value of the curvature or of the radius of curvature happens to be positive; but, if it happens to be negative, the contrary is true.

3.

If we refer the position of a point in the plane to two perpendicular axes of coordinates to which correspond the directions 0 and 90°, in such a manner that the first coordinate represents the distance of the point from the second axis, measured in the direction of the first axis; whereas the second coordinate represents the distance from the first axis, measured in the direction of the second axis; if, further, the indeterminates x, y represent the coordinates of a point on the curved line, s the length of the line measured from an arbitrary origin to this point, ϕ the direction of the tangent at this point, and r the radius of curvature; then we shall have

$$dx = \cos\phi \,.\, ds,$$
$$dy = \sin\phi \,.\, ds,$$
$$r = \frac{ds}{d\phi}.$$

If the nature of the curved line is defined by the equation $V = 0$, where V is a function of x, y, and if we set

$$d\,V = p\,dx + q\,dy,$$

then on the curved line

$$p\,dx + q\,dy = 0.$$

Hence

$$p\cos\phi + q\sin\phi = 0,$$

and therefore

$$\tan \phi = -\frac{p}{q}.$$

We have also

$$\cos \phi \,.\, dp + \sin \phi \,.\, dq - (p \sin \phi - q \cos \phi)\, d\phi = 0.$$

If, therefore, we set, according to a well known theorem,

$$dp = P\,dx + Q\,dy,$$
$$dq = Q\,dx + R\,dy,$$

then we have [2]

$$(P \cos^2 \phi + 2\, Q \cos \phi \sin \phi + R \sin^2 \phi)\, ds = (p \sin \phi - q \cos \phi)\, d\phi,$$

therefore

$$\frac{1}{r} = \frac{P \cos^2 \phi + 2\, Q \cos \phi \sin \phi + R \sin^2 \phi}{p \sin \phi - q \cos \phi},$$

or, since [3]

$$\cos \phi = \frac{\mp q}{\sqrt{(p^2 + q^2)}}, \qquad \sin \phi = \frac{\pm p}{\sqrt{(p^2 + q^2)}};$$

$$\pm \frac{1}{r} = \frac{P q^2 - 2\, Q p q + R p^2}{(p^2 + q^2)^{3/2}}.$$

4.

The ambiguous sign in the last formula might at first seem out of place, but upon closer consideration it is found to be quite in order. In fact, since this expression depends simply upon the partial differentials of V, and since the function V itself merely defines the nature of the curve without at the same time fixing the sense in which it is supposed to be described, the question, whether the curve is convex toward the right or left, must remain undetermined until the sense is determined by some other means. The case is similar in the determination of ϕ by means of the tangent, to single values of which correspond two angles differing by 180°. The sense in which the curve is described can be specified in the following different ways.

I. By means of the sign of the change in x. If x increases, then $\cos \phi$ must be positive. Hence the upper signs will hold if q has a negative value, and the lower signs if q has a positive value. When x decreases, the contrary is true.

II. By means of the sign of the change in y. If y increases, the upper signs must be taken when p is positive, the lower when p is negative. The contrary is true when y decreases.

III. By means of the sign of the value which the function V takes for points not on the curve. Let δx, δy be the variations of x, y when we go out from the curve toward the right, at right angles to the tangent, that is, in the direction $\phi + 90°$; and let the length of this normal be $\delta\rho$. Then, evidently, we have

$$\delta x = \delta\rho \,.\, \cos(\phi + 90°),$$
$$\delta y = \delta\rho \,.\, \sin(\phi + 90°),$$

or

$$\delta x = -\delta\rho \,.\, \sin\phi,$$
$$\delta y = +\delta\rho \,.\, \cos\phi.$$

Since now, when $\delta\rho$ is infinitely small,

$$\begin{aligned}\delta V &= p\,\delta x + q\,\delta y \\ &= (-p\sin\phi + q\cos\phi)\,\delta\rho \\ &= \mp\,\delta\rho\,\sqrt{(p^2+q^2)}\end{aligned}$$

and since on the curve itself V vanishes, the upper signs will hold if V, on passing through the curve from left to right, changes from positive to negative, and the contrary. If we combine this with what is said at the end of Art. 2, it follows that the curve is always convex toward that side on which V receives the same sign as

$$Pq^2 - 2\,Qpq + Rp^2.$$

For example, if the curve is a circle, and if we set

$$V = x^2 + y^2 - a^2$$

then we have

$$p = 2\,x, \qquad q = 2\,y,$$
$$P = 2, \qquad Q = 0, \qquad R = 2,$$
$$Pq^2 - 2\,Qpq + Rp^2 = 8\,y^2 + 8\,x^2 = 8\,a^2,$$
$$(p^2+q^2)^{3/2} = 8\,a^3,$$
$$r = \pm\,a$$

and the curve will be convex toward that side for which

$$x^2 + y^2 > a^2,$$

as it should be.

The side toward which the curve is convex, or, what is the same thing, the signs in the above formulæ, will remain unchanged by moving along the curve, so long as

$$\frac{\delta V}{\delta\rho}$$

does not change its sign. Since V is a continuous function, such a change can take place only when this ratio passes through the value zero. But this necessarily presupposes that p and q become zero at the same time. At such a point the radius of curvature becomes infinite or the curvature vanishes. Then, generally speaking, since here

$$-p\sin\phi+q\cos\phi$$

will change its sign, we have here a point of inflexion.

5.

The case where the nature of the curve is expressed by setting y equal to a given function of x, namely, $y=X$, is included in the foregoing, if we set

$$V=X-y.$$

If we put

$$dX=X'\,dx, \qquad dX'=X''\,dx,$$

then we have

$$p=X', \qquad q=-1,$$
$$P=X'', \qquad Q=0, \qquad R=0,$$

therefore

$$\pm\frac{1}{r}=\frac{X''}{(1+X'^2)^{3/2}}.$$

Since q is negative here, the upper sign holds for increasing values of x. We can therefore say, briefly, that for a positive X'' the curve is concave toward the same side toward which the y-axis lies with reference to the x-axis; while for a negative X'' the curve is convex toward this side.

6.

If we regard x, y as functions of s, these formulæ become still more elegant. Let us set

$$\frac{dx}{ds}=x', \qquad \frac{dx'}{ds}=x'',$$
$$\frac{dy}{ds}=y', \qquad \frac{dy'}{ds}=y''.$$

Then we shall have

$$x'=\cos\phi, \qquad y'=\sin\phi,$$
$$x''=-\frac{\sin\phi}{r}, \qquad y''=\frac{\cos\phi}{r};$$

or

$$y'=-r\,x'', \qquad x'=r\,y'',$$

or also

$$1 = r\,(x'\,y'' - y'\,x''),$$

so that

$$x'\,y'' - y'\,x''$$

represents the curvature, and

$$\frac{1}{x'\,y'' - y'\,x''}$$

the radius of curvature.

7.

We shall now proceed to the consideration of curved surfaces. In order to represent the directions of straight lines in space considered in its three dimensions, we imagine a sphere of unit radius described about an arbitrary centre. Accordingly, a point on this sphere will represent the direction of all straight lines parallel to the radius whose extremity is at this point. As the positions of all points in space are determined by the perpendicular distances x, y, z from three mutually perpendicular planes, the directions of the three principal axes, which are normal to these principal planes, shall be represented on the auxiliary sphere by the three points (1), (2), (3). These points are, therefore, always 90° apart, and at once indicate the sense in which the coordinates are supposed to increase. We shall here state several well known theorems, of which constant use will be made.

1) The angle between two intersecting straight lines is measured by the arc [of the great circle] between the points on the sphere which represent their directions.

2) The orientation of every plane can be represented on the sphere by means of the great circle in which the sphere is cut by the plane through the centre parallel to the first plane.

3) The angle between two planes is equal to the angle between the great circles which represent their orientations, and is therefore also measured by the angle between the poles of the great circles.

4) If x, y, z; x', y', z' are the coordinates of two points, r the distance between them, and L the point on the sphere which represents the direction of the straight line drawn from the first point to the second, then

$$x' = x + r\cos(1)L,$$
$$y' = y + r\cos(2)L,$$
$$z' = z + r\cos(3)L.$$

5) It follows immediately from this that we always have

$$\cos^2(1)L + \cos^2(2)L + \cos^2(3)L = 1$$

[and] also, if L' is any other point on the sphere,

$$\cos(1)L \,.\, \cos(1)L' + \cos(2)L \,.\, \cos(2)L' + \cos(3)L \,.\, \cos(3)L' = \cos LL'.$$

We shall add here another theorem, which has appeared nowhere else, as far as we know, and which can often be used with advantage.

Let L, L', L'', L''' be four points on the sphere, and A the angle which LL''' and $L'L''$ make at their point of intersection. [Then we have]

$$\cos LL' \,.\, \cos L''L''' - \cos LL'' \,.\, \cos L'L''' = \sin LL''' \,.\, \sin L'L'' \,.\, \cos A.$$

The proof is easily obtained in the following way. Let

$$AL = t, \qquad AL' = t', \qquad AL'' = t'', \qquad AL''' = t''';$$

we have then

$$\begin{aligned}
\cos LL' &= \cos t \cos t' + \sin t \sin t' \cos A,\\
\cos L''L''' &= \cos t'' \cos t''' + \sin t'' \sin t''' \cos A,\\
\cos LL'' &= \cos t \cos t'' + \sin t \sin t'' \cos A,\\
\cos L'L''' &= \cos t' \cos t''' + \sin t' \sin t''' \cos A.
\end{aligned}$$

Therefore

$$\begin{aligned}
&\cos LL' \cos L''L''' - \cos LL'' \cos L'L''\\
&\quad = \cos A \,\{\cos t \cos t' \sin t'' \sin t''' + \cos t'' \cos t''' \sin t \sin t'\\
&\qquad\qquad - \cos t \cos t'' \sin t' \sin t''' - \cos t' \cos t''' \sin t \sin t''\}\\
&\quad = \cos A \,(\cos t \sin t''' - \cos t''' \sin t)\,(\cos t' \sin t'' - \cos t'' \sin t')\\
&\quad = \cos A \sin (t''' - t) \sin (t'' - t')\\
&\quad = \cos A \sin LL''' \sin L'L''.
\end{aligned}$$

Since each of the two great circles goes out from A in two opposite directions, two supplementary angles are formed at this point. But it is seen from our analysis that those branches must be chosen, which go in the same sense from L toward L''' and from L' toward L''.

Instead of the angle A, we can take also the distance of the pole of the great circle LL''' from the pole of the great circle $L'L''$. However, since every great circle has two poles, we see that we must join those about which the great circles run in the same sense from L toward L''' and from L' toward L'', respectively.

The development of the special case, where one or both of the arcs LL''' and $L'L''$ are 90°, we leave to the reader.

6) Another useful theorem is obtained from the following analysis. Let L, L', L'' be three points upon the sphere and put

$$\cos L\,(1) = x, \quad \cos L\,(2) = y, \quad \cos L\,(3) = z,$$
$$\cos L'\,(1) = x', \quad \cos L'\,(2) = y', \quad \cos L'\,(3) = z',$$
$$\cos L''(1) = x'', \quad \cos L''(2) = y'', \quad \cos L''(3) = z''.$$

We assume that the points are so arranged that they run around the triangle included by them in the same sense as the points (1), (2), (3). Further, let λ be that pole of the great circle $L'L''$ which lies on the same side as L. We then have, from the above lemma,

$$y'z'' - z'y'' = \sin L'L'' . \cos \lambda(1),$$
$$z'x'' - x'z'' = \sin L'L'' . \cos \lambda(2),$$
$$x'y'' - y'x'' = \sin L'L'' . \cos \lambda(3).$$

Therefore, if we multiply these equations by x, y, z respectively, and add the products, we obtain

$$xy'z'' + x'y''z + x''yz' - xy''z' - x'yz'' - x''y'z = \sin L'L'' . \cos \lambda L,$$ [4]

wherefore, we can write also, according to well known principles of spherical trigonometry,[5]

$$\sin L'L'' . \sin L\,L'' . \sin L'$$
$$= \sin L'L'' . \sin L\,L' . \sin L''$$
$$= \sin L'L'' . \sin L'L'' . \sin L,$$

if L, L', L'' denote the three angles of the spherical triangle. At the same time we easily see that this value is one-sixth of the pyramid whose angular points are the centre of the sphere and the three points L, L', L'' (and indeed *positive*, if etc.). [4]

8.

The nature of a curved surface is defined by an equation between the coordinates of its points, which we represent by

$$f(x, y, z) = 0.$$

Let the total differential of $f(x, y, z)$ be

$$P\,dx + Q\,dy + R\,dz,$$

where P, Q, R are functions of x, y, z. We shall always distinguish two sides of the surface, one of which we shall call the upper, and the other the lower. Generally speaking, on passing through the surface the value of f changes its sign, so that, as long as the continuity is not interrupted, the values are positive on one side and negative on the other.

The direction of the normal to the surface toward that side which we regard as the upper side is represented upon the auxiliary sphere by the point L. Let

$$\cos L(1) = X, \qquad \cos L(2) = Y, \qquad \cos L(3) = Z.$$

Also let ds denote an infinitely small line upon the surface; and, as its direction is denoted by the point λ on the sphere, let

$$\cos\lambda(1) = \xi, \qquad \cos\lambda(2) = \eta, \qquad \cos\lambda(3) = \zeta.$$

We then have

$$dx = \xi\, ds, \qquad dy = \eta\, ds, \qquad dz = \zeta\, ds,$$

therefore

$$P\xi + Q\eta + R\zeta = 0,$$

and, since λL must be equal to 90°, we have also

$$X\xi + Y\eta + Z\zeta = 0.$$

Since P, Q, R, X, Y, Z depend only on the position of the surface on which we take the element, and since these equations hold for every direction of the element on the surface, it is easily seen that P, Q, R must be proportional to X, Y, Z. Therefore

$$P = X\mu, \qquad Q = Y\mu, \qquad R = Z\mu,$$

Therefore, since

$$X^2 + Y^2 + Z^2 = 1;$$
$$\mu = PX + QY + RZ$$

and

$$\mu^2 = P^2 + Q^2 + R^2,$$

or

$$\mu = \pm\sqrt{(P^2 + Q^2 + R^2)}.$$

If we go out from the surface, in the direction of the normal, a distance equal to the element $\delta\rho$, then we shall have

$$\delta x = X\delta\rho, \qquad \delta y = Y\delta\rho, \qquad \delta z = Z\delta\rho$$

and

$$\delta f = P\,\delta x + Q\,\delta y + R\,\delta z = \mu\,\delta\rho.$$ [6]

We see, therefore, how the sign of μ depends on the change of sign of the value of f in passing from the lower to the upper side.

9.

Let us cut the curved surface by a plane through the point to which our notation refers; then we obtain a plane curve of which ds is an element, in connection with which we shall retain the above notation. We shall regard as the upper side of the plane that one on which the normal to the curved surface lies. Upon this plane we erect a normal whose direction is expressed by the point $\mathfrak{L}$ of the auxiliary sphere. By moving along the curved line, λ and L will therefore change their positions, while $\mathfrak{L}$ remains constant, and λL and $\lambda \mathfrak{L}$ are always equal to 90°. Therefore λ describes the great circle one of whose poles is $\mathfrak{L}$. The element of this great circle will be equal to $\frac{ds}{r}$, if r denotes the radius of curvature of the curve. And again, if we denote the direction of this element upon the sphere by λ', then λ' will evidently lie in the same great circle and be 90° from λ as well as from $\mathfrak{L}$. If we now set

$$\cos \lambda'(1) = \xi', \qquad \cos \lambda'(2) = \eta', \qquad \cos \lambda'(3) = \zeta',$$

then we shall have

$$d\xi = \xi' \frac{ds}{r}, \qquad d\eta = \eta' \frac{ds}{r}, \qquad d\zeta = \zeta' \frac{ds}{r},$$

since, in fact, ξ, η, ζ are merely the coordinates of the point λ referred to the centre of the sphere.

Since by the solution of the equation $f(x, y, z) = 0$ the coordinate z may be expressed in the form of a function of x, y, we shall, for greater simplicity, assume that this has been done and that we have found

$$z = F(x, y).$$

We can then write as the equation of the surface

$$z - F(x, y) = 0,$$

or

$$f(x, y, z) = z - F(x, y).$$

From this follows, if we set

$$dF(x, y) = t\,dx + u\,dy,$$

$$P = -t, \qquad Q = -u, \qquad R = 1,$$

where t, u are merely functions of x and y. We set also

$$dt = T\,dx + U\,dy, \qquad du = U\,dx + V\,dy.$$

Therefore upon the whole surface we have

$$dz = t\,dx + u\,dy$$

and therefore, on the curve,

$$\zeta = t\xi + u\eta.$$

Hence differentiation gives, on substituting the above values for $d\xi$, $d\eta$, $d\zeta$,

$$(\zeta' - t\xi' - u\eta') \frac{ds}{r} = \xi\,dt + \eta\,du$$
$$= (\xi^2 T + 2\,\xi\eta U + \eta^2 V)\,ds,$$

or

$$\frac{1}{r} = \frac{\xi^2 T + 2\,\xi\eta U + \eta^2 V}{-\xi' t - \eta' \mu + \zeta'}$$
$$= \frac{Z(\xi^2 T + 2\,\xi\eta U + \eta^2 V)}{X\xi' + Y\eta' + Z\zeta'}$$
$$= \frac{Z(\xi^2 T + 2\,\xi\eta U + \eta^2 V)}{\cos L\lambda'}.$$

10.

Before we further transform the expression just found, we will make a few remarks about it.

A normal to a curve in its plane corresponds to two directions upon the sphere, according as we draw it on the one or the other side of the curve. The one direction, toward which the curve is *concave*, is denoted by λ', the other by the opposite point on the sphere. Both these points, like L and $\mathfrak{L}$, are 90° from λ, and therefore lie in a great circle. And since $\mathfrak{L}$ is also 90° from λ, $\mathfrak{L}L = 90° - L\lambda'$, or $= L\lambda' - 90°$. Therefore

$$\cos L\lambda' = \pm \sin \mathfrak{L}L,$$

where $\sin \mathfrak{L}L$ is necessarily positive. Since r is regarded as positive in our analysis, the sign of $\cos L\lambda'$ will be the same as that of

$$Z(\xi^2 T + 2\,\xi\eta\, U + \eta^2 V).$$

And therefore a positive value of this last expression means that $L\lambda'$ is less than 90°, or that the curve is concave toward the side on which lies the projection of the normal to the surface upon the plane. A negative value, on the contrary, shows that the curve is convex toward this side. Therefore, in general, we may set also

$$\frac{1}{r} = \frac{Z(\xi^2 T + 2\,\xi\eta\, U + \eta^2 V)}{\sin \mathfrak{L}L},$$

if we regard the radius of curvature as positive in the first case, and negative in the second. $\mathfrak{L}L$ is here the angle which our cutting plane makes with the plane tangent to the curved surface, and we see that in the different cutting planes passed through the same point and the same tangent the radii of curvature are proportional to the sine of the inclination. Because of this simple relation, we shall limit ourselves hereafter to the case where this angle is a right angle, and where the cutting plane, therefore, is passed through the normal of the curved surface. Hence we have for the radius of curvature the simple formula

$$\frac{1}{r} = Z(\xi^2 T + 2\,\xi\eta U + \eta^2 V).$$

11.

Since an infinite number of planes may be passed through this normal, it follows that there may be infinitely many different values of the radius of curvature. In this case T, U, V, Z are regarded as constant, ξ, η, ζ as variable. In order to make the latter depend upon a single variable, we take two fixed points M, M' $90°$ apart on the great circle whose pole is L. Let their coordinates referred to the centre of the sphere be $\alpha, \beta, \gamma;\ \alpha', \beta', \gamma'$. We have then

$$\cos\lambda(1) = \cos\lambda M \,.\, \cos M(1) + \cos\lambda M' \,.\, \cos M'(1) + \cos\lambda L \,.\, \cos L(1).$$

If we set

$$\lambda M = \phi,$$

then we have

$$\cos\lambda M' = \sin\phi,$$

and the formula becomes

$$\xi = \alpha\cos\phi + \alpha'\sin\phi,$$

and likewise

$$\eta = \beta\cos\phi + \beta'\sin\phi,$$
$$\zeta = \gamma\cos\phi + \gamma'\sin\phi.$$

Therefore, if we set

$$A = (\alpha^2 T + 2\,\alpha\beta U + \beta^2 V)Z,$$
$$B = (\alpha\alpha' T + (\alpha'\beta + \alpha\beta')U + \beta\beta' V)Z,$$ [7]
$$C = (\alpha'^2 T + 2\,\alpha'\beta' U + \beta'^2 V)Z,$$

we shall have

$$\frac{1}{r} = A\cos^2\phi + 2\,B\cos\phi\sin\phi + C\sin^2\phi$$
$$= \frac{A+C}{2} + \frac{A-C}{2}\cos 2\,\phi + B\sin 2\,\phi.$$

If we put

$$\frac{A-C}{2} = E\cos 2\,\theta,$$
$$B = E\sin 2\,\theta,$$

where we may assume that E has the same sign as $\frac{A-C}{2}$, then we have

$$\frac{1}{r}=\tfrac{1}{2}(A+C)+E\cos 2(\phi-\theta).$$

It is evident that ϕ denotes the angle between the cutting plane and another plane through this normal and that tangent which corresponds to the direction M. Evidently, therefore, $\frac{1}{r}$ takes its greatest (absolute) value, or r its smallest, when $\phi=\theta$; and $\frac{1}{r}$ its smallest absolute value, when $\phi=\theta+90°$. Therefore the greatest and the least curvatures occur in two planes perpendicular to each other. Hence these extreme values for $\frac{1}{r}$ are

$$\tfrac{1}{2}(A+C)\pm\sqrt{\left\{\left(\frac{A-C}{2}\right)^2+B^2\right\}}.$$

Their sum is $A+C$ and their product is $AC-B^2$, or the product of the two extreme radii of curvature is

$$=\frac{1}{AC-B^2}.$$

This product, which is of great importance, merits a more rigorous development. In fact, from formulæ above we find

$$AC-B^2=(\alpha\beta'-\beta\alpha')^2(TV-U^2)Z^2.$$

But from the third formula in [Theorem] 6, Art. 7, we easily infer that [8]

$$\alpha\beta'-\beta\alpha'=\pm Z,$$

therefore

$$AC-B^2=Z^4(TV-U^2).$$

Besides, from Art. 8,

$$Z=\pm\frac{R}{\sqrt{(P^2+Q^2+R^2)}}$$

$$=\pm\frac{1}{\sqrt{(1+t^2+u^2)}},$$

therefore

$$AC-B^2=\frac{TV-U^2}{(1+t^2+u^2)^2}.$$

Just as to *each* point on the curved surface corresponds a particular point L on the auxiliary sphere, by means of the normal erected at this point and the radius of the auxiliary sphere parallel to the normal, so the aggregate of the points on the auxiliary sphere, which correspond to all the points of a *line* on the curved surface, forms a line which will correspond to the line on the curved surface. And, likewise, to every finite figure on the curved surface will correspond a finite figure on the auxiliary sphere, the area of which upon the latter shall be regarded as the measure of the amplitude of the former. We shall either regard this area as a number, in which case the square of the radius of the auxiliary sphere is the unit, or else express it in degrees, etc., setting the area of the hemisphere equal to 360°.

The comparison of the area upon the curved surface with the corresponding amplitude leads to the idea of what we call the measure of curvature of the surface. If the former is proportional to the latter, the curvature is called uniform; and the quotient, when we divide the amplitude by the surface, is called the measure of curvature. This is the case when the curved surface is a sphere, and the measure of curvature is then a fraction whose numerator is unity and whose denominator is the square of the radius.

We shall regard the measure of curvature as positive, if the boundaries of the figures upon the curved surface and upon the auxiliary sphere run in the same sense; as negative, if the boundaries enclose the figures in contrary senses. If they are not proportional, the surface is non-uniformily curved. And at each point there exists a particular measure of curvature, which is obtained from the comparison of corresponding infinitesimal parts upon the curved surface and the auxiliary sphere. Let $d\sigma$ be a surface element on the former, and $d\Sigma$ the corresponding element upon the auxiliary sphere, then

$$\frac{d\Sigma}{d\sigma}$$

will be the measure of curvature at this point.

In order to determine their boundaries, we first project both upon the xy-plane. The magnitudes of these projections are $Z\,d\sigma,\,Z\,d\Sigma$. The sign of Z will show whether the boundaries run in the same sense or in contrary senses around the surfaces and their projections. We will suppose that the figure is a triangle; the projection upon the xy-plane has the coordinates

$$x, y; \qquad x + dx,\, y + dy; \qquad x + \delta x,\, y + \delta y.$$

Hence its double area will be

$$2\,Z\,d\sigma = dx\,.\,\delta y - dy\,.\,\delta x.$$

To the projection of the corresponding element upon the sphere will correspond the coordinates:

$$
\begin{array}{ll}
X, & Y, \\
X+\frac{\partial X}{\partial x}\cdot dx+\frac{\partial X}{\partial y}\cdot dy, & Y+\frac{\partial Y}{\partial x}\cdot dx+\frac{\partial Y}{\partial y}\cdot dy, \\
X+\frac{\partial X}{\partial x}\cdot \delta x+\frac{\partial X}{\partial y}\cdot \delta y, & Y+\frac{\partial Y}{\partial x}\cdot \delta x+\frac{\partial Y}{\partial y}\cdot \delta y.
\end{array}
$$

From this the double area of the element is found to be

$$
\begin{aligned}
2\,Z\,d\Sigma = & \left(\frac{\partial X}{\partial x}\cdot dx+\frac{\partial X}{\partial y}\cdot dy\right)\left(\frac{\partial Y}{\partial x}\cdot \delta x+\frac{\partial Y}{\partial y}\cdot \delta y\right) \\
& -\left(\frac{\partial X}{\partial x}\cdot \delta x+\frac{\partial X}{\partial y}\cdot \delta y\right)\left(\frac{\partial Y}{\partial x}\cdot dx+\frac{\partial Y}{\partial y}\cdot dy\right) \\
= & \left(\frac{\partial X}{\partial x}\cdot\frac{\partial Y}{\partial y}-\frac{\partial X}{\partial y}\cdot\frac{\partial Y}{\partial x}\right)(dx\,.\,\delta y-dy\,.\,\delta x).
\end{aligned}
$$

The measure of curvature is, therefore,

$$
=\frac{\partial X}{\partial x}\cdot\frac{\partial Y}{\partial y}-\frac{\partial X}{\partial y}\cdot\frac{\partial Y}{\partial x}=\omega.
$$

Since

$$
X=-t\,Z, \qquad Y=-u\,Z,
$$
$$
(1+t^2+u^2)\,Z^2=1,
$$

we have

$$
\begin{aligned}
dX &= -Z^3(1+u^2)\,dt+Z^3\,tu\,.\,du, \\
dY &= +Z^3tu\,.\,dt-Z^3(1+t^2)\,du,
\end{aligned}
$$

therefore

$$
\begin{array}{ll}
\frac{\partial X}{\partial x}=Z^3\{-(1+u^2)T+tuU\}, & \frac{\partial Y}{\partial x}=Z^3\{tu\,T-(1+t^2)U\}, \\
\frac{\partial X}{\partial y}=Z^3\{-(1+u^2)U+tuV\}, & \frac{\partial Y}{\partial y}=Z^3\{tu\,U-(1+t^2)\,V\},
\end{array}
$$

and

$$
\begin{aligned}
\omega &= Z^6(TV-U^2)\big((1+t^2)(1+u^2)-t^2u^2\big) \\
&= Z^6(TV-U^2)(1+t^2+u^2) \\
&= Z^4(TV-U^2) \\
&= \frac{TV-U^2}{(1+t^2+u^2)^2},
\end{aligned}
$$

the very same expression which we have found at the end of the preceding article. Therefore we see that

"The measure of curvature is always expressed by means of a fraction whose numerator is unity and whose denominator is the product of the maximum and minimum radii of curvature in the planes passing through the normal."

12.

We will now investigate the nature of shortest lines upon curved surfaces. The nature of a curved line in space is determined, in general, in such a way that the coordinates x, y, z of each point are regarded as functions of a single variable, which we shall call w. The length of the curve, measured from an arbitrary origin to this point, is then equal to

$$\int \sqrt{\left\{\left(\frac{dx}{dw}\right)^2 + \left(\frac{dy}{dw}\right)^2 + \left(\frac{dz}{dw}\right)^2\right\}} \cdot dw.$$

If we allow the curve to change its position by an infinitely small variation, the variation of the whole length will then be

$$= \int \frac{\frac{dx}{dw} \cdot d\,\delta x + \frac{dy}{dw} \cdot d\,\delta y + \frac{dz}{dw} \cdot d\,\delta z}{\sqrt{\left\{\left(\frac{dx}{dw}\right)^2 + \left(\frac{dy}{dw}\right)^2 + \left(\frac{dz}{dw}\right)^2\right\}}} = \frac{\frac{dx}{dw} \cdot \delta x + \frac{dy}{dw} \cdot \delta y + \frac{dz}{dw} \cdot \delta z}{\sqrt{\left\{\left(\frac{dx}{dw}\right)^2 + \left(\frac{dy}{dw}\right)^2 + \left(\frac{dz}{dw}\right)^2\right\}}}$$

$$- \int \left\{ \delta x \,.\, d \frac{\frac{dx}{dw}}{\sqrt{\left\{\left(\frac{dx}{dw}\right)^2 + \left(\frac{dy}{dw}\right)^2 + \left(\frac{dz}{dw}\right)^2\right\}}} + \delta y \,.\, d \frac{\frac{dy}{dw}}{\sqrt{\left\{\left(\frac{dx}{dw}\right)^2 + \left(\frac{dy}{dw}\right)^2 + \left(\frac{dz}{dw}\right)^2\right\}}} \right.$$

$$\left. + \, \delta z \,.\, d \frac{\frac{dz}{dw}}{\sqrt{\left\{\left(\frac{dx}{dw}\right)^2 + \left(\frac{dy}{dw}\right)^2 + \left(\frac{dz}{dw}\right)^2\right\}}} \right\}.$$

The expression under the integral sign must vanish in the case of a minimum, as we know. Since the curved line lies upon a given curved surface whose equation is

$$P\,dx + Q\,dy + R\,dz = 0,$$

the equation between the variations δx, δy, δz

$$P\,\delta x + Q\,\delta y + R\,\delta z = 0$$

must also hold. From this, by means of well known principles, we easily conclude that the differentials

$$d \cdot \frac{\frac{dx}{dw}}{\sqrt{\left\{\left(\frac{dx}{dw}\right)^2 + \left(\frac{dy}{dw}\right)^2 + \left(\frac{dz}{dw}\right)^2\right\}}}, \quad d \cdot \frac{\frac{dy}{dw}}{\sqrt{\left\{\left(\frac{dx}{dw}\right)^2 + \left(\frac{dy}{dw}\right)^2 + \left(\frac{dz}{dw}\right)^2\right\}}},$$

$$d \cdot \frac{\frac{dz}{dw}}{\sqrt{\left\{\left(\frac{dx}{dw}\right)^2 + \left(\frac{dy}{dw}\right)^2 + \left(\frac{dz}{dw}\right)^2\right\}}}$$

must be proportional to the quantities P, Q, R respectively. If ds is an element of the curve; λ the point upon the auxiliary sphere, which represents the direction of this element; L the point giving the direction of the normal as above; and ξ, η, ζ; X, Y, Z the coordinates of the points λ, L referred to the centre of the auxiliary sphere, then we have

$$dx = \xi\, ds, \qquad dy = \eta\, ds, \qquad dz = \zeta\, ds,$$
$$\xi^2 + \eta^2 + \zeta^2 = 1.$$

Therefore we see that the above differentials will be equal to $d\xi, d\eta, d\zeta$. And since P, Q, R are proportional to the quantities X, Y, Z, the character of the shortest line is such that

$$\frac{d\xi}{X} = \frac{d\eta}{Y} = \frac{d\zeta}{Z}.$$

13.

To every point of a curved line upon a curved surface there correspond two points on the sphere, according to our point of view; namely, the point λ, which represents the direction of the linear element, and the point L, which represents the direction of the normal to the surface. The two are evidently 90° apart. In our former investigation (Art. 9), where [we] supposed the curved line to lie in a plane, we had *two* other points upon the sphere; namely, $\mathfrak{L}$, which represents the direction of the normal to the plane, and λ', which represents the direction of the normal to the element of the curve in the plane. In this case, therefore, $\mathfrak{L}$ was a fixed point and λ, λ' were always in a great circle whose pole was $\mathfrak{L}$. In generalizing these considerations, we shall retain the notation $\mathfrak{L}, \lambda'$, but we must define the meaning of these symbols from a more general point of view. When the curve s is described, the points L, λ also describe curved lines upon the auxiliary sphere, which, generally speaking, are no longer great circles. Parallel to the element of the second line, we draw a radius of the auxiliary sphere to the point λ', but instead of this point we take the point opposite when λ' is more than 90° from L. In the first case, we regard the element at λ as positive, and in the other as negative. Finally, let $\mathfrak{L}$ be the point on the auxiliary sphere, which is 90° from both λ and λ', and which is so taken that λ, λ', $\mathfrak{L}$ lie in the same order as (1), (2), (3).

The coordinates of the four points of the auxiliary sphere, referred to its centre, are for

L	X	Y	Z
λ	ξ	η	ζ
λ'	ξ'	η'	ζ'
$\mathfrak{L}$	α	β	γ.

Hence each of these 4 points describes a line upon the auxiliary sphere, whose elements we shall express by dL, $d\lambda$, $d\lambda'$, $d\mathfrak{L}$. We have, therefore,

$$d\xi = \xi' d\lambda,$$
$$d\eta = \eta' d\lambda,$$
$$d\zeta = \zeta' d\lambda.$$

In an analogous way we now call

$$\frac{d\lambda}{ds}$$

the measure of curvature of the curved line upon the curved surface, and its reciprocal

$$\frac{ds}{d\lambda}$$

the radius of curvature. If we denote the latter by ρ, then

$$\rho\, d\xi = \xi' ds,$$
$$\rho\, d\eta = \eta' ds,$$
$$\rho\, d\zeta = \zeta' ds.$$

If, therefore, our line be a shortest line, ξ', η', ζ' must be proportional to the quantities X, Y, Z. But, since at the same time

$$\xi'^2 + \eta'^2 + \zeta'^2 = X^2 + Y^2 + Z^2 = 1,$$

we have

$$\xi' = \pm X, \quad \eta' = \pm Y, \quad \zeta' = \pm Z,$$

and since, further,

$$\begin{aligned} \xi' X + \eta' Y + \zeta' Z &= \cos \lambda' L \\ &= \pm (X^2 + Y^2 + Z^2) \\ &= \pm 1, \end{aligned}$$

and since we always choose the point λ' so that

$$\lambda' L < 90^\circ,$$

then for the shortest line

$$\lambda' L = 0,$$

or λ' and L must coincide. Therefore

$$\rho\, d\xi = X\, ds,$$
$$\rho\, d\eta = Y\, ds,$$
$$\rho\, d\zeta = Z\, ds,$$

and we have here, instead of 4 curved lines upon the auxiliary sphere, only 3 to consider. Every element of the second line is therefore to be regarded as lying in the great circle $L\lambda$. And the positive or negative value of ρ refers to the concavity or the convexity of the curve in the direction of the normal.

14.

We shall now investigate the spherical angle upon the auxiliary sphere, which the great circle going from L toward λ makes with that one going from L toward one of the fixed points (1), (2), (3); *e. g.*, toward (3). In order to have something definite here, we shall consider the sense from $L(3)$ to $L\lambda$ the same as that in which (1), (2), and (3) lie. If we call this angle ϕ, then it follows from the theorem of Art. 7 that

$$\sin L(3) \,.\, \sin L\lambda \,.\, \sin\phi = Y\xi - X\eta, \quad 9$$

or, since $L\lambda = 90^\circ$ and

$$\sin L(3) = \sqrt{(X^2 + Y^2)} = \sqrt{(1 - Z^2)},$$

we have

$$\sin\phi = \frac{Y\xi - X\eta}{\sqrt{(X^2 + Y^2)}}.$$

Furthermore,

$$\sin L(3) \,.\, \sin L\lambda \,.\, \cos\phi = \zeta,$$

or

$$\cos\phi = \frac{\zeta}{\sqrt{(X^2 + Y^2)}}$$

and

$$\tan\phi = \frac{Y\xi - X\eta}{\zeta} = \frac{\zeta'}{\zeta}.$$

Hence we have

$$d\phi = \frac{\zeta Y d\xi - \zeta X d\eta - (Y\xi - X\eta)d\zeta + \xi\zeta dY - \eta\zeta dX}{(Y\xi - X\eta)^2 + \zeta^2}.$$

The denominator of this expression is

$$\begin{aligned} &= Y^2\xi^2 - 2XY\xi\eta + X^2\eta^2 + \zeta^2 \\ &= -(X\xi + Y\eta)^2 + (X^2 + Y^2)(\xi^2 + \eta^2) + \zeta^2 \\ &= -Z^2\zeta^2 + (1 - Z^2)(1 - \zeta^2) + \zeta^2 \\ &= 1 - Z^2, \end{aligned}$$

or

$$d\phi = \frac{\zeta Y d\xi - \zeta X d\eta + (X\eta - Y\xi)d\zeta - \eta\zeta dX + \xi\zeta dY}{1 - Z^2}.$$

We verify readily by expansion the identical equation

$$\begin{aligned} &\eta\zeta(X^2 + Y^2 + Z^2) + YZ(\xi^2 + \eta^2 + \zeta^2) \\ &= (X\xi + Y\eta + Z\zeta)(Z\eta + Y\zeta) + (X\zeta - Z\xi)(X\eta - Y\xi) \end{aligned}$$

and likewise

$$\begin{aligned} &\xi\zeta(X^2 + Y^2 + Z^2) + XZ(\xi^2 + \eta^2 + \zeta^2) \\ &= (X\xi + Y\eta + Z\zeta)(X\zeta + Z\xi) + (Y\xi - X\eta)(Y\zeta - Z\eta). \end{aligned}$$

We have, therefore,

$$\begin{aligned} \eta\zeta &= -YZ + (X\zeta - Z\xi)(X\eta - Y\xi), \\ \xi\zeta &= -XZ + (Y\xi - X\eta)(Y\zeta - Z\eta). \end{aligned}$$

Substituting these values, we obtain

$$\begin{aligned} d\phi = &\frac{Z}{1 - Z^2}(YdX - XdY) + \frac{\zeta Y d\xi - \zeta X d\eta}{1 - Z^2} \\ &+ \frac{X\eta - Y\xi}{1 - Z^2}\{d\zeta - (X\zeta - Z\xi)dX - (Y\zeta - Z\eta)dY\}. \end{aligned}$$

Now

$$\begin{aligned} XdX + YdY + ZdZ &= 0, \\ \xi dX + \eta dY + \zeta dZ &= -Xd\xi - Yd\eta - Zd\zeta. \end{aligned}$$

On substituting we obtain, instead of what stands in the parenthesis,

$$d\zeta - Z(Xd\xi + Yd\eta + Zd\zeta).$$

Hence

$$\begin{aligned} d\phi = &\frac{Z}{1 - Z^2}(YdX - XdY) + \frac{d\xi}{1 - Z^2}\{\zeta Y - \eta X^2 Z + \xi XYZ\} \\ &- \frac{d\eta}{1 - Z^2}\{\zeta X + \eta XYZ - \xi Y^2 Z\} \quad 10 \\ &+ d\zeta(\eta X - \xi Y). \end{aligned}$$

Since, further,

$$\eta X^2 Z - \xi XYZ = \eta X^2 Z + \eta Y^2 Z + \zeta ZYZ$$
$$= \eta Z(1 - Z^2) + \zeta YZ^2,$$
$$\eta XYZ - \xi Y^2 Z = -\xi X^2 Z - \zeta XZ^2 - \xi Y^2 Z$$
$$= -\xi Z(1 - Z^2) - \zeta XZ^2,$$

our whole expression becomes

$$d\phi = \frac{Z}{1-Z^2}(YdX - XdY)$$
$$+ (\zeta Y - \eta Z)d\xi + (\xi Z - \zeta X)d\eta + (\eta X - \xi Y)d\zeta.$$

15.

The formula just found is true in general, whatever be the nature of the curve. But if this be a shortest line, then it is clear that the last three terms destroy each other, and consequently

$$d\phi = -\frac{Z}{1-Z^2}(XdY - YdX).$$

But we see at once that [11]

$$\frac{Z}{1-Z^2}(XdY - YdX)$$

is nothing but the area of the part of the auxiliary sphere, which is formed between the element of the line L, the two great circles drawn through its extremities and

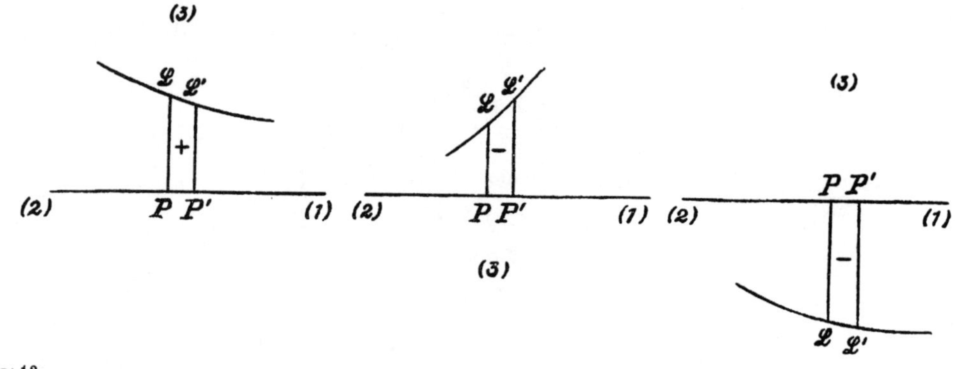

(3),[12] and the element thus intercepted on the great circle through (1) and (2). This surface is considered positive, if L and (3) lie on the same side of (1) (2), and if the

direction from P to P' is the same as that from (2) to (1); negative, if the contrary of one of these conditions hold; positive again, if the contrary of both conditions be true. In other words, the surface is considered positive if we go around the circumference of the figure $LL'P'P$ in the same sense as (1) (2) (3); negative, if we go in the contrary sense.

If we consider now a finite part of the line from L to L' and denote by ϕ, ϕ' the values of the angles at the two extremities, then we have

$$\phi' = \phi + \text{Area } LL'P'P,$$

the sign of the area being taken as explained.[13]

Now let us assume further that,[14] from the origin upon the curved surface, infinitely many other shortest lines go out, and denote by A that indefinite angle which the first element, moving counter-clockwise, makes with the first element of the first line; and through the other extremities of the different curved lines let a curved line be drawn, concerning which, first of all, we leave it undecided whether it be a shortest line or not. If we suppose also that those indefinite values, which for the first line were ϕ, ϕ', be denoted by ψ, ψ' for each of these lines, then $\psi' - \psi$ is capable of being represented in the same manner on the auxiliary sphere by the space $LL'_1P'_1P$.[15] Since evidently $\psi = \phi - A$, the space

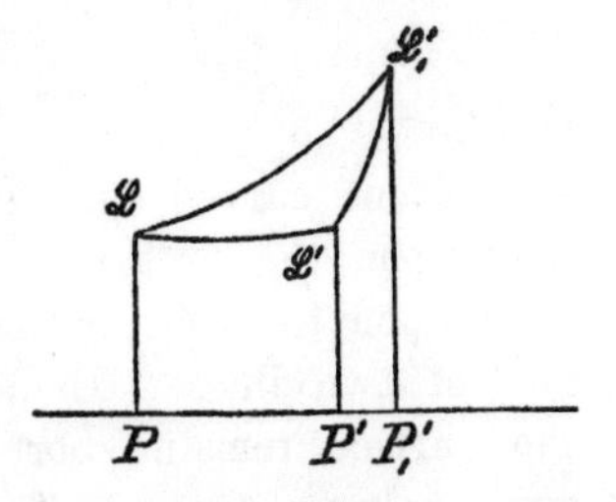

$$\begin{aligned} LL'_1P'_1P'L'L &= \psi' - \psi - \phi' + \phi \\ &= \psi' - \phi' + A \\ &= LL'_1L'L + L'L'_1P'_1P'. \end{aligned}$$

If the bounding line is also a shortest line, and, when prolonged, makes with LL', LL'_1 the angles B, B_1; if, further, χ, χ_1 denote the same at the points L', L'_1, that ϕ did at L in the line LL', then we have

$$\chi_1 = \chi + \text{Area } L'L'_1P'_1P',$$
$$\psi' - \phi' + A = LL'_1L'L + \chi_1 - \chi;$$

but

$$\phi' = \chi + B,$$
$$\psi' = \chi_1 + B_1,$$

therefore

$$B_1 - B + A = LL'_1L'L.$$

The angles of the triangle $LL'L'_1$ evidently are

$$A, \qquad 180° - B, \qquad B_1,$$

therefore their sum is

$$180^\circ + LL'_1L'L.$$

The form of the proof will require some modification and explanation, if the point (3) falls within the triangle. But, in general, we conclude

> "The sum of the three angles of a triangle, which is formed of shortest lines upon an arbitrary curved surface, is equal to the sum of 180° and the area of the triangle upon the anxiliary sphere, the boundary of which is formed by the points L, corresponding to the points in the boundary of the original triangle, and in such a manner that the area of the triangle may be regarded as positive or negative according as it is inclosed by its boundary in the same sense as the original figure or the contrary."

Wherefore we easily conclude also that the sum of all the angles of a polygon of n sides, which are shortest lines upon the curved surface, is [equal to] the sum of $(n-2)\ 180^\circ +$ the area of the polygon upon the sphere etc.[16]

16.

If one curved surface can be completely developed upon another surface, then all lines upon the first surface will evidently retain their magnitudes after the development upon the other surface; likewise the angles which are formed by the intersection of two lines. Evidently, therefore, such lines also as are shortest lines upon one surface remain shortest lines after the development. Whence, if to any arbitrary polygon formed of shortest lines, while it is upon the first surface, there corresponds the figure of the zeniths[17] upon the auxiliary sphere, the area of which is A, and if, on the other hand, there corresponds to the same polygon, after its development upon another surface, a figure of the zeniths upon the auxiliary sphere, the area of which is A', it follows at once that in every case

$$A = A'.$$

Although this proof originally presupposes the boundaries of the figures to be shortest lines, still it is easily seen that it holds generally, whatever the boundary may be. For, in fact, if the theorem is independent of the number of sides, nothing will prevent us from imagining for every polygon, of which some or all of its sides are not shortest lines, another of infinitely many sides all of which are shortest lines.

Further, it is clear that every figure retains also its area after the transformation by development.

We shall here consider 4 figures:

1) an arbitrary figure upon the first surface,
2) the figure on the auxiliary sphere, which corresponds to the zeniths of the previous figure,
3) the figure upon the second surface, which No. 1 forms by the development,
4) the figure upon the auxiliary sphere, which corresponds to the zeniths of No. 3.

Therefore, according to what we have proved, 2 and 4 have equal areas, as also 1 and 3. Since we assume these figures infinitely small, the quotient obtained by dividing 2 by 1 is the measure of curvature of the first curved surface at this point, and likewise the quotient obtained by dividing 4 by 3, that of the second surface. From this follows the important theorem:

"In the transformation of surfaces by development the measure of curvature at every point remains unchanged."

This is true, therefore, of the product of the greatest and smallest radii of curvature.

In the case of the plane, the measure of curvature is evidently everywhere zero. Whence follows therefore the important theorem:

"For all surfaces developable upon a plane the measure of curvature everywhere vanishes,"

or

$$\left(\frac{\partial^2 z}{\partial x \partial y}\right)^2 - \left(\frac{\partial^2 z}{\partial x^2}\right)\left(\frac{\partial^2 z}{\partial y^2}\right) = 0,$$

which criterion is elsewhere derived from other principles, though, as it seems to us, not with the desired rigor. It is clear that in all such surfaces the zeniths of all points can not fill out any space, and therefore they must all lie in a line.

17.

From a given point on a curved surface we shall let an infinite number of shortest lines go out, which shall be distinguished from one another by the angle which their first elements make with the first element of a *definite* shortest line. This angle we shall call θ. Further, let s be the length [measured from the given point] of a part of such a shortest line, and let its extremity have the coordinates x, y, z. Since θ and s, therefore, belong to a perfectly definite point on the curved surface, we can regard x, y, z as functions of θ and s. The direction of the element of s corresponds to the point λ on the sphere, whose coordinates are ξ, η, ζ. Thus we shall have

$$\xi=\frac{\partial x}{\partial s}, \qquad \eta=\frac{\partial y}{\partial s}, \qquad \zeta=\frac{\partial z}{\partial s}.$$

The extremities of all shortest lines of equal lengths s correspond to a curved line whose length we may call t. We can evidently consider t as a function of s and θ, and if the direction of the element of t corresponds upon the sphere to the point λ' whose coordinates are ξ', η', ζ', we shall have

$$\xi'\cdot\frac{\partial t}{\partial\theta}=\frac{\partial x}{\partial\theta}, \qquad \eta'\cdot\frac{\partial t}{\partial\theta}=\frac{\partial y}{\partial\theta}, \qquad \zeta'\cdot\frac{\partial t}{\partial\theta}=\frac{\partial z}{\partial\theta}.$$

Consequently

$$(\xi\xi'+\eta\eta'+\zeta\zeta')\frac{\partial t}{\partial\theta}=\frac{\partial x}{\partial s}\cdot\frac{\partial x}{\partial\theta}+\frac{\partial y}{\partial s}\cdot\frac{\partial y}{\partial\theta}+\frac{\partial z}{\partial s}\cdot\frac{\partial z}{\partial\theta}.$$

This magnitude we shall denote by u, which itself, therefore, will be a function of θ and s.

We find, then, if we differentiate with respect to s,

$$\frac{\partial u}{\partial s}=\frac{\partial^2 x}{\partial s^2}\cdot\frac{\partial x}{\partial\theta}+\frac{\partial^2 y}{\partial s^2}\cdot\frac{\partial y}{\partial\theta}+\frac{\partial^2 z}{\partial s^2}\cdot\frac{\partial z}{\partial\theta}+\frac{1}{2}\,\frac{\partial\left\{\left(\frac{\partial x}{\partial s}\right)^2+\left(\frac{\partial y}{\partial s}\right)^2+\left(\frac{\partial z}{\partial s}\right)^2\right\}}{\partial\theta}$$

$$=\frac{\partial^2 x}{\partial s^2}\cdot\frac{\partial x}{\partial\theta}+\frac{\partial^2 y}{\partial s^2}\cdot\frac{\partial y}{\partial\theta}+\frac{\partial^2 z}{\partial s^2}\cdot\frac{\partial z}{\partial\theta},$$

because

$$\left(\frac{\partial x}{\partial s}\right)^2+\left(\frac{\partial y}{\partial s}\right)^2+\left(\frac{\partial z}{\partial s}\right)^2=1,$$

and therefore its differential is equal to zero.

But since all points [belonging] to one constant value of θ lie on a shortest line, if we denote by L the zenith of the point to which s, θ correspond and by X, Y, Z the coordinates of L, [from the last formulæ of Art. 13],

$$\frac{\partial^2 x}{\partial s^2}=\frac{X}{p}, \qquad \frac{\partial^2 y}{\partial s^2}=\frac{Y}{p}, \qquad \frac{\partial^2 z}{\partial s^2}=\frac{Z}{p},$$

if p is the radius of curvature. We have, therefore,

$$p\cdot\frac{\partial u}{\partial s}=X\frac{\partial x}{\partial\theta}+Y\frac{\partial y}{\partial\theta}+Z\frac{\partial z}{\partial\theta}=\frac{\partial t}{\partial\theta}(X\xi'+Y\eta'+Z\zeta').$$

But

$$X\xi'+Y\eta'+Z\zeta'=\cos L\lambda'=0,$$

because, evidently, λ' lies on the great circle whose pole is L. Therefore we have

$$\frac{\partial u}{\partial s}=0,$$

or u independent of s, and therefore a function of θ alone. But for $s=0$, it is evident that $t=0$, $\frac{\partial t}{\partial \theta}=0$, and therefore $u=0$. Whence we conclude that, in general, $u=0$, or

$$\cos \lambda\lambda' = 0.$$

From this follows the beautiful theorem:

> "If all lines drawn from a point on the curved surface are shortest lines of equal lengths, they meet the line which joins their extremities everywhere at right angles."

We can show in a similar manner that, if upon the curved surface any curved line whatever is given, and if we suppose drawn from every point of this line toward the same side of it and at right angles to it only shortest lines of equal lengths, the extremities of which are joined by a line, this line will be cut at right angles by those lines in all its points. We need only let θ in the above development represent the length of the *given* curved line from an arbitrary point, and then the above calculations retain their validity, except that $u=0$ for $s=0$ is now contained in the hypothesis.

18.

The relations arising from these constructions deserve to be developed still more fully. We have, in the first place, if, for brevity, we write m for $\frac{\partial t}{\partial \theta}$,

$$(1)\qquad \frac{\partial x}{\partial s}=\xi, \quad \frac{\partial y}{\partial s}=\eta, \quad \frac{\partial z}{\partial s}=\zeta,$$

$$(2)\qquad \frac{\partial x}{\partial \theta}=m\xi', \quad \frac{\partial y}{\partial \theta}=m\eta', \quad \frac{\partial z}{\partial \theta}=m\zeta',$$

$$(3)\qquad \xi^2+\eta^2+\zeta^2=1,$$

$$(4)\qquad \xi'^2+\eta'^2+\zeta'^2=1,$$

$$(5)\qquad \xi\xi'+\eta\eta'+\zeta\zeta'=0.$$

Furthermore,

$$(6)\qquad X^2+Y^2+Z^2=1,$$

$$(7)\qquad X\xi+Y\eta+Z\zeta=0,$$

$$(8)\qquad X\xi'+Y\eta'+Z\zeta'=0,$$

and

$$[9]\qquad \begin{cases} X=\zeta\eta'-\eta\zeta', \\ Y=\xi\zeta'-\zeta\xi', \\ Z=\eta\xi'-\xi\eta'; \end{cases}$$

[10]
$$\begin{cases} \xi' = \eta Z - \zeta Y, \\ \eta' = \zeta X - \xi Z, \\ \zeta' = \xi Y - \eta X; \end{cases}$$

[11]
$$\begin{cases} \xi = Y\zeta' - Z\eta', \\ \eta = Z\xi' - X\zeta', \\ \zeta = X\eta' - Y\xi'. \end{cases}$$

Likewise, $\frac{\partial\xi}{\partial s}, \frac{\partial\eta}{\partial s}, \frac{\partial\zeta}{\partial s}$ are proportional to X, Y, Z, and if we set

$$\frac{\partial\xi}{\partial s} = pX, \qquad \frac{\partial\eta}{\partial s} = pY, \qquad \frac{\partial\zeta}{\partial s} = pZ,$$

where $\frac{1}{p}$ denotes the radius of curvature of the line s, then

$$p = X\frac{\partial\xi}{\partial s} + Y\frac{\partial\eta}{\partial s} + Z\frac{\partial\zeta}{\partial s}.$$

By differentiating (7) with respect to s, we obtain

$$-p = \xi\frac{\partial X}{\partial s} + \eta\frac{\partial Y}{\partial s} + \zeta\frac{\partial Z}{\partial s}.$$

We can easily show that $\frac{\partial\xi'}{\partial s}, \frac{\partial\eta'}{\partial s}, \frac{d\zeta'}{\partial s}$ also are proportional to X, Y, Z. In fact, [from 10] the values of these quantities are also [equal to]

$$\eta\frac{\partial Z}{\partial s} - \zeta\frac{\partial Y}{\partial s}, \qquad \zeta\frac{\partial X}{\partial s} - \xi\frac{\partial Z}{\partial s}, \qquad \xi\frac{\partial Y}{\partial s} - \eta\frac{\partial X}{\partial s},$$

therefore

$$\begin{aligned} Y\frac{\partial\xi'}{\partial s} - X\frac{\partial\eta'}{\partial s} &= -\zeta\left(\frac{Y\partial Y}{\partial s} + \frac{X\partial X}{\partial s}\right) + \frac{\partial Z}{\partial s}(Y\eta + X\xi) \\ &= -\zeta\left(\frac{X\partial X + Y\partial Y + Z\partial Z}{\partial s}\right) + \frac{\partial Z}{\partial s}(X\xi + Y\eta + Z\zeta) \\ &= 0, \end{aligned}$$

and likewise the others. We set, therefore,

$$\frac{\partial\xi'}{\partial s} = p'X, \qquad \frac{\partial\eta'}{\partial s} = p'Y, \qquad \frac{\partial\zeta'}{\partial s} = p'Z,$$

whence

$$p' = \pm\sqrt{\left\{\left(\frac{\partial\xi'}{\partial s}\right)^2 + \left(\frac{\partial\eta'}{\partial s}\right)^2 + \left(\frac{\partial\zeta'}{\partial s}\right)^2\right\}}$$

and also

$$p' = X\frac{\partial \xi'}{\partial s} + Y\frac{\partial \eta'}{\partial s} + Z\frac{\partial \zeta'}{\partial s}.$$

Further [we obtain], from the result obtained by differentiating (8),

$$-p' = \xi'\frac{\partial X}{\partial s} + \eta'\frac{\partial Y}{\partial s} + \zeta'\frac{\partial Z}{\partial s}.$$

But we can derive two other expressions for this. We have

$$\frac{\partial m\xi'}{\partial s} = \frac{\partial \xi}{\partial \theta}, \qquad \left[\frac{\partial m\eta'}{\partial s} = \frac{\partial \eta}{\partial \theta}, \qquad \frac{\partial m\zeta'}{\partial s} = \frac{\partial \zeta}{\partial \theta},\right]$$

therefore [because of (8)]

$$mp' = X\frac{\partial \xi}{\partial \theta} + Y\frac{\partial \eta}{\partial \theta} + Z\frac{\partial \zeta}{\partial \theta},$$

[and therefore, from (7),]

$$-mp' = \xi\frac{\partial X}{\partial \theta} + \eta\frac{\partial Y}{\partial \theta} + \zeta\frac{\partial Z}{\partial \theta}.$$

After these preliminaries [using (2) and (4)] we shall now first put m in the form

$$m = \xi'\frac{\partial x}{\partial \theta} + \eta'\frac{\partial y}{\partial \theta} + \zeta'\frac{\partial z}{\partial \theta},$$

and differentiating with respect to s, we have*

$$\begin{aligned}\frac{\partial m}{\partial s} &= \frac{\partial x}{\partial \theta}\cdot\frac{\partial \xi'}{\partial s} + \frac{\partial y}{\partial \theta}\cdot\frac{\partial \eta'}{\partial s} + \frac{\partial z}{\partial \theta}\cdot\frac{\partial \zeta'}{\partial s}\\ &\quad + \xi'\frac{\partial^2 x}{\partial s.\partial \theta} + \eta'\frac{\partial^2 y}{\partial s.\partial \theta} + \zeta'\frac{\partial^2 z}{\partial s.\partial \theta}\\ &= mp'(\xi'X + \eta'Y + \zeta'Z)\\ &\quad + \xi'\frac{\partial \xi}{\partial \theta} + \eta'\frac{\partial \eta}{\partial \theta} + \zeta'\frac{\partial \zeta}{\partial \theta}\\ &= \xi'\frac{\partial \xi}{\partial \theta} + \eta'\frac{\partial \eta}{\partial \theta} + \zeta'\frac{\partial \zeta}{\partial \theta}.\end{aligned}$$

---

* It is better to differentiate m^2. [In fact from (2) and (4)

$$m^2 = \left(\frac{\partial x}{\partial \theta}\right)^2 + \left(\frac{\partial y}{\partial \theta}\right)^2 + \left(\frac{\partial z}{\partial \theta}\right)^2,$$

therefore

$$\begin{aligned}m\frac{\partial m}{\partial s} &= \frac{\partial x}{\partial \theta}\cdot\frac{\partial^2 x}{\partial \theta\partial s} + \frac{\partial y}{\partial \theta}\cdot\frac{\partial^2 y}{\partial \theta\partial s} + \frac{\partial z}{\partial \theta}\cdot\frac{\partial^2 z}{\partial \theta\partial s}\\ &= m\xi'\frac{\partial \xi}{\partial \theta} + m\eta'\frac{\partial \eta}{\partial \theta} + m\zeta'\frac{\partial \zeta}{\partial \theta}.\Big]\end{aligned}$$

If we differentiate again with respect to s, and notice that

$$\frac{\partial^2 \xi}{\partial s \partial \theta} = \frac{\partial (pX)}{\partial \theta}, \text{ etc.,}$$

and that

$$X\xi' + Y\eta' + Z\zeta' = 0,$$

we have

$$\begin{aligned}
\frac{\partial^2 m}{\partial s^2} &= p\left(\xi' \frac{\partial X}{\partial \theta} + \eta' \frac{\partial Y}{\partial \theta} + \zeta' \frac{\partial Z}{\partial \theta}\right) + p'\left(X \frac{\partial \xi}{\partial \theta} + Y \frac{\partial \eta}{\partial \theta} + Z \frac{\partial \zeta}{\partial \theta}\right) \\
&= p\left(\xi' \frac{\partial X}{\partial \theta} + \eta' \frac{\partial Y}{\partial \theta} + \zeta' \frac{\partial Z}{\partial \theta}\right) + m p'^2 \\
&= -\left(\xi \frac{\partial X}{\partial s} + \eta \frac{\partial Y}{\partial s} + \zeta \frac{\partial Z}{\partial s}\right)\left(\xi' \frac{\partial X}{\partial \theta} + \eta' \frac{\partial Y}{\partial \theta} + \zeta' \frac{\partial Z}{\partial \theta}\right) \\
&\quad + \left(\xi' \frac{\partial X}{\partial s} + \eta' \frac{\partial Y}{\partial s} + \zeta' \frac{\partial Z}{\partial s}\right)\left(\xi \frac{\partial X}{\partial \theta} + \eta \frac{\partial Y}{\partial \theta} + \zeta \frac{\partial Z}{\partial \theta}\right) \\
&= \left(\frac{\partial Y}{\partial \theta}\frac{\partial Z}{\partial s} - \frac{\partial Y}{\partial s}\frac{\partial Z}{\partial \theta}\right)X + \left(\frac{\partial Z}{\partial \theta}\frac{\partial X}{\partial s} - \frac{\partial Z}{\partial s}\frac{\partial X}{\partial \theta}\right)Y + \left(\frac{\partial X}{\partial \theta}\frac{\partial Y}{\partial s} - \frac{\partial X}{\partial s}\frac{\partial Y}{\partial \theta}\right)Z. \quad {}^{18}
\end{aligned}$$

[But if the surface element

$$m\, ds\, d\theta$$

belonging to the point x, y, z be represented upon the auxiliary sphere of unit radius by means of parallel normals, then there corresponds to it an area whose magnitude is

$$\left\{ X\left(\frac{\partial Y}{\partial s}\frac{\partial Z}{\partial \theta} - \frac{\partial Y}{\partial \theta}\frac{\partial Z}{\partial s}\right) + Y\left(\frac{\partial Z}{\partial s}\frac{\partial X}{\partial \theta} - \frac{\partial Z}{\partial \theta}\frac{\partial X}{\partial s}\right) + Z\left(\frac{\partial X}{\partial s}\frac{\partial Y}{\partial \theta} - \frac{\partial X}{\partial \theta}\frac{\partial Y}{\partial s}\right) \right\} ds\, d\theta.$$

Consequently, the measure of curvature at the point under consideration is equal to

$$-\frac{1}{m}\frac{\partial^2 m}{\partial s^2}.]$$

NOTES.

1. The parts enclosed in brackets are additions of the editor of the German edition or of the translators.

"The foregoing fragment, *Neue allgemeine Untersuchungen über die krummen Flächen*, differs from the *Disquisitiones* not only in the more limited scope of the matter, but also in the method of treatment and the arrangement of the theorems. There [paper of 1827] GAUSS assumes that the rectangular coordinates x, y, z of a point of the surface can be expressed as functions of any two independent variables p and q, while here [paper of 1825] he chooses as new variables the geodesic coordinates s and θ. Here [paper of 1825] he begins by proving the theorem, that the sum of the three angles of a triangle, which is formed by shortest lines upon an arbitrary curved surface, differs from 180° by the area of the triangle, which corresponds to it in the representation by means of parallel normals upon the auxiliary sphere of unit radius. From this, by means of simple geometrical considerations, he derives the fundamental theorem, that "in the transformation of surfaces by bending, the measure of curvature at every point remains unchanged." But there [paper of 1827] he first shows, in Art. 11, that the measure of curvature can be expressed simply by means of the three quantities E, F, G, and their derivatives with respect to p and q, from which follows the theorem concerning the invariant property of the measure of curvature as a corollary; and only much later, in Art. 20, quite independently of this, does he prove the theorem concerning the sum of the angles of a geodesic triangle."

Remark by Stäckel, Gauss's Works, vol. VIII, p. 443.

2. Art. 3, p. 84, l. 9. $\cos^2\phi$, etc., is used here where the German text has $\cos\phi^2$, etc.

3. Art. 3, p. 84, l. 13. p^2, etc., is used here where the German text has pp, etc.

4. Art. 7, p. 89, ll. 13, 21. Since λL is less than 90°, $\cos\lambda L$ is always positive and, therefore, the algebraic sign of the expression for the volume of this pyramid depends upon that of $\sin L'L''$. Hence it is positive, zero, or negative according as the arc $L'L''$ is less than, equal to, or greater than 180°.

5. Art. 7, p. 89, ll. 14–21. As is seen from the paper of 1827 (see page 6), Gauss

corrected this statement. To be correct it should read: for which we can write also, according to well known principles of spherical trigonometry,

$$\sin LL' \,.\, \sin L' \,.\, \sin L'L'' = \sin L'L'' \,.\, \sin L'' \,.\, \sin L''L = \sin L''L \,.\, \sin L \,.\, \sin LL',$$

if L, L', L'' denote the three angles of the spherical triangle, where L is the angle measured from the arc LL'' to LL', and so for the other angles. At the same time we easily see that this value is one-sixth of the pyramid whose angular points are the centre of the sphere and the three points L, L', L''; and this pyramid is *positive* when the points L, L', L'' are arranged in the same order about this triangle as the points (1), (2), (3) about the triangle (1) (2) (3).

6. Art. 8, p. 90, l. 7 fr. bot. In the German text V stands for f in this equation and in the next line but one.

7. Art. 11, p. 93, l. 8 fr. bot. In the German text, in the expression for B, $(\alpha\beta' + \alpha\beta')$ stands for $(\alpha'\beta + \alpha\beta')$.

8. Art. 11, p. 94, l. 17. The vertices of the triangle are M, M', (3), whose coordinates are α, β, γ; α', β', γ'; 0, 0, 1, respectively. The pole of the arc MM' on the same side as (3) is L, whose coordinates are X, Y, Z. Now applying the formula on page 89, line 10,

$$x'y'' - y'x'' = \sin L'L'' \cos \lambda(3),$$

to this triangle, we obtain

$$\alpha\beta' - \beta\alpha' = \sin MM' \cos L(3)$$

or, since

$$MM' = 90^\circ, \text{ and } \cos L(3) = \pm Z$$

we have

$$\alpha\beta' - \beta\alpha' = \pm Z.$$

9. Art. 14, p. 100, l. 19. Here X, Y, Z; ξ, η, ζ; 0, 0, 1 take the place of x, y, z; x', y', z'; x'', y'', z'' of the top of page 89. Also (3), λ take the place of L', L'', and ϕ is the angle L in the note at the top of this page.

10. Art. 14, p. 101, l. 2 fr. bot. In the German text $\{\zeta X - \eta XYZ + \xi Y^2 Z\}$ stands for $\{\zeta X + \eta XYZ - \xi Y^2 Z\}$.

11. Art. 15, p. 102, l. 13 and the following. Transforming to polar coordinates, r, θ, ψ, by the substitutions (since on the auxiliary sphere $r = 1$)

$$X = \sin\theta \sin\psi, \qquad Y = \sin\theta\cos\psi, \qquad Z = \cos\theta,$$

$$dX = \sin\theta\cos\psi\, d\psi + \cos\theta\sin\psi\, d\theta, \qquad dY = -\sin\theta\sin\psi\, d\psi + \cos\theta\cos\psi\, d\theta,$$

$$(1) \qquad = \frac{Z}{1 - Z^2}(X\,dY - Y\,dX) \qquad \text{becomes} \qquad \cos\theta\, d\psi.$$

In the figures on page 102, PL and $P'L'$ are arcs of great circles intersecting in the point (3), and the element LL', which is not necessarily the arc of a great circle, corresponds to the element of the geodesic line on the curved surface. $(2)PP'(1)$ also is the arc of a great circle. Here $P'P=d\psi$, $Z=\cos\theta=$ Altitude of the zone of which $LL'P'P$ is a part. The area of a zone varies as the altitude of the zone. Therefore, in the case under consideration,

$$\frac{\text{Area of zone}}{2\pi}=\frac{Z}{1}.$$

Also

$$\frac{\text{Area } LL'P'P}{\text{Area of zone}}=\frac{d\psi}{2\pi}.$$

From these two equations,

$$\text{(2)} \qquad \text{Area } LL'P'P=Z\,d\psi, \text{ or } \cos\theta\, d\psi.$$

From (1) and (2)

$$-\frac{Z}{1-Z^2}(XdY-YdX)=\text{Area } LL'P'P.$$

12. Art. 15, p. 102. The point (3) in the figures on this page was added by the translators.

13. Art. 15, p. 103, ll. 6–9. It has been shown that $d\phi=$ Area $LL'P'P,=dA$, say. Then

$$\int_{\phi}^{\phi'} d\phi=\int_{0}^{A} dA,$$

or

$$\phi'-\phi=A, \text{ the finite area } LL'P'P.$$

14. Art. 15, p. 103, l. 10 and the following. Let A, B', B_1 be the vertices of a geodesic triangle on the curved surface, and let the corresponding triangle on the auxiliary sphere be $LL'L'_1L$, whose sides are not necessarily arcs of great circles. Let A, B', B_1 denote also the angles of the geodesic triangle. Here B' is the supplement of the angle denoted by B on page 103. Let ϕ be the angle on the sphere between the great circle arcs $L\lambda, L(3)$, *i. e.*, $\phi=(3)L\lambda$, λ corresponding to the direction of the element at A on the geodesic line AB', and let $\phi'=(3)L'\lambda_1$, λ_1 corresponding to the direction of the element at B' on the line AB'. Similarly, let $\psi=(3)L\mu$,

$\psi' = (3)L'_1\mu_1$, μ, μ_1 denoting the directions of the elements at A, B_1, respectively, on the line AB_1. And let $\chi = (3)L'\nu$, $\chi_1 = (3)L'_1\nu_1$, ν, ν_1 denoting the directions of the elements at B', B_1, respectively, on the line $B'B_1$.

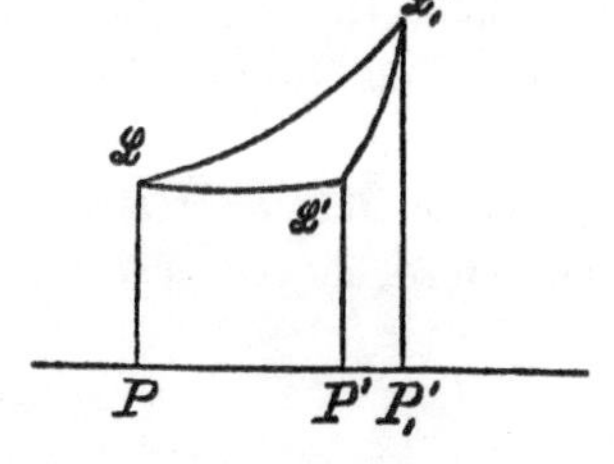

Then from the first formula on page 103,

$$\phi' - \phi = \text{Area } LL'P'P,$$
$$\psi' - \psi = \text{Area } LL'_1P'_1P,$$
$$\chi_1 - \chi = \text{Area } L'L'_1P'_1P',$$

$$\psi' - \psi - (\phi' - \phi) - (\chi_1 - \chi) = \text{Area } LL'_1P'_1P - \text{Area } LL'P'P - \text{Area } L'L'_1P'_1P',$$

or

(1) $$(\phi - \psi) + (\chi - \phi') + (\psi' - \chi_1) = \text{Area } LL'_1L'L.$$

Since λ, μ represent the directions of the linear elements at A on the geodesic lines AB', AB_1, respectively, the absolute value of the angle A on the surface is measured by the arc $\mu\lambda$, or by the spherical angle $\mu L\lambda$. But $\phi - \psi = (3)L\lambda - (3)L\mu = \mu L\lambda$.

Therefore

$$A = \phi - \psi.$$

Similarly

$$180° - B' = -(\chi - \phi'),$$
$$B_1 = \psi' - \chi_1.$$

Therefore, from (1),

$$A + B' + B_1 - 180° = \text{Area } LL'_1L'L.$$

15. Art. 15, p. 103, l. 19. In the German text $LL'P'P$ stands for $LL'_1P'_1P$, which represents the angle $\psi' - \psi$.

16. Art. 15, p. 104, l. 12. This general theorem may be stated as follows:

The sum of all the angles of a polygon of n sides, which are shortest lines upon the curved surface, is equal to the sum of $(n-2)180°$ and the area of the polygon upon the auxiliary sphere whose boundary is formed by the points L which correspond to the points of the boundary of the given polygon, and in such a manner that the area of this polygon may be regarded positive or negative according as it is enclosed by its boundary in the same sense as the given figure or the contrary.

17. Art. 16, p. 104, l. 12 fr. bot. The *zenith* of a point on the surface is the corresponding point on the auxiliary sphere. It is the spherical representation of the point.

18. Art. 18, p. 110, l. 10. The normal to the surface is here taken in the direction opposite to that given by [9] page 107.

[Line numbers are counted from the top of the page, except those marked b, which are counted from the bottom.]

p. 3, §1 [As the introduction discusses, the practical rationale for considering "the directions of various straight lines in space" comes from astronomy and surveying. Gauss explains in his abstract (p. 45) that his "auxiliary sphere" of unit radius is derived from the use of the celestial sphere in astronomy: any directed line A can be represented by the coordinates of a point on the celestial sphere whose radius vector is parallel to A:

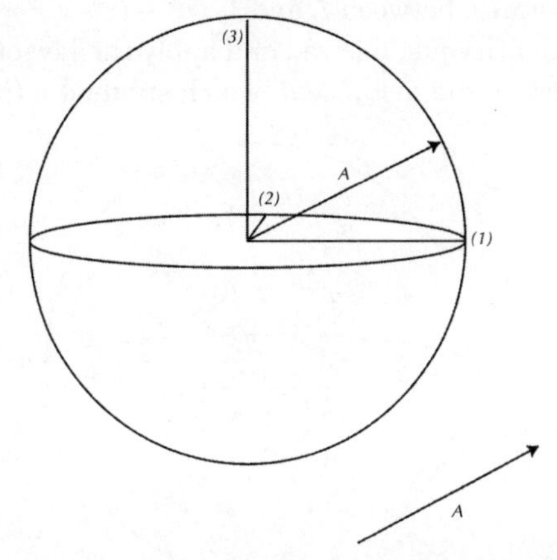

Note that by (1), (2), (3) Gauss indicates orthogonal ("quadrant," as he calls them) directions (corresponding to the usual x, y, z coordinates) with respect to which he will measure angles.]

p. 3, §2I [Thus, if at radius r the angle θ measures arc length s, then $s = r\theta$.]

p. 3, §2II [A *great circle* is a circle on the surface of a sphere such that the plane defined by that circle passes through the center of the sphere. A great circle always cuts its sphere into two equal parts.]

p. 4, §2III [A *spherical angle* is the angle between two intersecting arcs of great circles of a sphere, measured by the plane angle formed by the tangents to the arcs at the point of intersection.]

p. 4, §2IV [The terms cos $(1)L$, cos $(2)L$, cos $(3)L$ are now referred to as "direction cosines";

the angles that Gauss calls (1)L, (2)L, (3)L are those measured with respect to the x, y, z axes respectively so that cos (1)$L = (x'-x)/r$, cos (2)$L = (y'-y)/r$, cos (3)$L = (z'-z)/r$ and the auxiliary sphere has unit radius $r = 1$.]

p. 4, §2V [The square of the distance between points x, y, z and x', y', z', using the definitions of §2IV, is $(x'-x)^2 + (y'-y)^2 + (z'-z)^2 = \cos^2$ (1)$L + \cos^2$ (2)$L + \cos^2$ (3)$L = 1$, because the points lie on the unit sphere, proving the first identity. To prove the second identity, take another point L' defined by $x'' = x' + r \cos$ (1)L', $y'' = y' + r \cos$ (2)L', $z'' = z' + r \cos$ (3)L'. Then consider the square of the distance between L and L', $d^2 = (x''-x')^2 + (y''-y')^2 + (z''-z')^2$. Expand this expression, use the identity just derived and apply the law of cosines in plane trigonometry to the triangle formed by the distance d, which subtend a (plane) angle LL' at the center of the sphere:

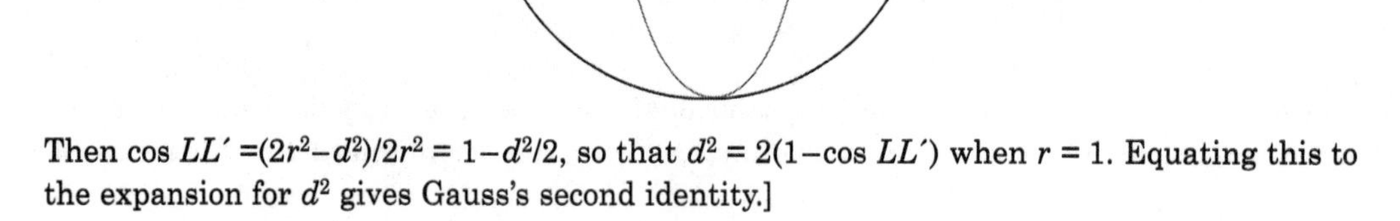

Then $\cos LL' = (2r^2-d^2)/2r^2 = 1-d^2/2$, so that $d^2 = 2(1-\cos LL')$ when $r = 1$. Equating this to the expansion for d^2 gives Gauss's second identity.]

p. 4, §2VI [Here is a diagram of the situation:

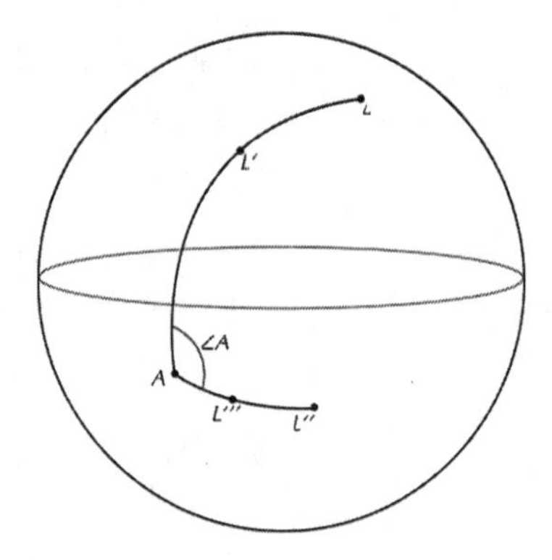

We need the *law of cosines in spherical trigonometry*, given in the appendix below (p. 123). Gauss applies this four times, each time on a particular spherical triangle made out of L, L', L'', L''', A (see also the figures on p. 51) to obtain the four ratios in lines 7-10. The final step is to multiply these as indicated and factor, as Gauss shows.]

p. 5, §2VII [Here is a diagram of the situation Gauss builds up:

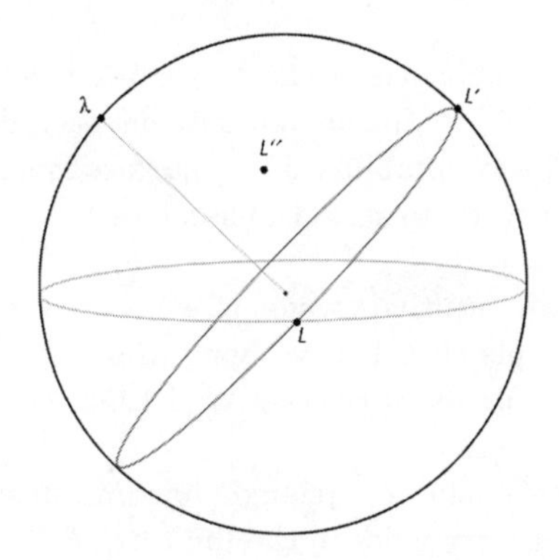

Note that λ is not necessarily on one of the axes but instead denotes the "zenith" point corresponding to the "horizon" defined by points L, L' (which Gauss places 90° apart), thinking of them in astronomical terms. To apply theorem §2VI to this case, note that in terms of the lettering used in that theorem, in the present case we need to substitute $L \to \lambda$, $L' \to L$, $L'' \to L'$, $L''' \to L''$, which then gives Gauss's first result: $yz' - y'z = \cos(1)\lambda \sin LL'$, where $(1)\lambda$ means (as above) the angle between direction (1) and point λ. The other two cases follow by analogous arguments; even more simply, they can be derived by permuting the names of the points cyclically: let $x \to y$, $y \to z$, $z \to x$, $(1) \to (2)$, $(2) \to (3)$, etc. In what follows, such cyclic permutation is often helpful to generate all other cases from one.]

p. 6, ll 1-5 [By "perpendicular" Gauss means a perpendicular to the line LL' drawn along a great circle. In l. 3, he states the *law of sines for spherical triangles*, given in the appendix below (p. 123). In l. 5, he uses the fact that λ is the "zenith" defined by the "horizon" LL', so that the angle $\lambda L'' \pm p = 90°$. He then uses the standard plane trigonometric identity for the sine of a sum or difference (here, of $p = 90° \mp \lambda L''$) and substitutes for the expression for Δ.]

p. 6, l. 4b [By *normal* Gauss means orthogonal or perpendicular.]

p. 7, l. 5 [Using the result from §2V.]

p. 7, l. 5b [This second method of *Gaussian coordinates* p, q is essential to Gauss's subsequent argument; compare the two latitude and longitude coordinates as a way of locating points on the earth's surface. Note also that in the equations at the bottom of this page a' is the coefficient of dq; the prime mark does not denote a derivative.]

p. 8, ll. 5-7 [These two linear equations are solved using Cramer's rule; note the use of the determinant Δ and its appropriate minors such as $bc' - cb$.]

p. 8, §6 [Note that here Gauss emphasizes that the surface is what now would be called an *orientable manifold*, having a definite inside and outside, each determined by the direction of the appropriate normals. The orientability of the surface is also an important element of the surfaces used in electrodynamics to state Gauss's Law.]

p. 11, l. 13 [Here Gauss gives an intuitive version of what later was rigorously proven as the *Jordan curve theorem*: any simple closed curve lying in a plane divides the plane into two exclusive parts, inside and outside the curve (see Wall 1993, 111-120).]

p. 12, ll. 1-2 [The key is that this spherical triangle has infinitesimal sides, which therefore can be approximated by ordinary triangles in the limit $dx, dy, \delta x, \delta y \to 0$:

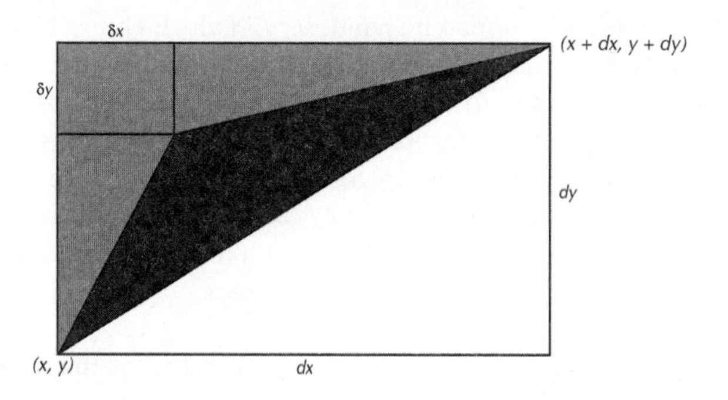

The area of the entire rectangle is $dx\,dy$, hence the area of the sought-for darkly-shaded triangle is half the entire rectangle, $\frac{1}{2}\,dx\,dy$, less the other parts (shown in lighter shading), namely $\frac{1}{2}\,dx\,dy - \delta x\,(dy - \delta y) - \frac{1}{2}\,\delta x\,\delta y - \frac{1}{2}\,(dx - \delta x)(dy - \delta y) = \frac{1}{2}\,dx\,\delta y - \frac{1}{2}\,\delta x\,dy$, as Gauss claims. An exactly parallel argument holds for the sphere, in the same limit.]

p. 12, l. 2b [Gauss means that these values have been substituted in the expression just given for k.]

p. 13, l. 7f [By "the formulæ given above," Gauss means the expressions for X, Y, Z according to the third method, given at the end of §4 on p. 8, ll. 9b, 11b. To get the following equations, Gauss takes the derivatives of each of these expressions for X, Y, Z; he also divides the

dZ expression on both sides by $2Z$ and uses the substitution $Z^2 = 1/(1 + t^2 + u^2)$. He then uses his expression $dZ = -Z^3(t\ dt + u\ du)$ to eliminate dZ throughout. To find the expressions given in ll. 10b-7b for $\partial X/\partial x$ etc., Gauss takes the partial derivatives of his expression for dX (in this case) and applies his definitions from the top of this page (ll. 2-4). The resultant expression for k is sometimes called the *Gauss equation*.]

p. 13, ll. 2b-1b [Gauss notes here that the curvature can be locally "swept away" through a suitable choice of coordinates in an infinitesimal region around a given point. In the case of gravitation, Einstein will make the same argument: in a freely-falling elevator, gravitation appears locally "swept away": objects in the elevator, in free fall, appear not to be subject to gravity. Note that, in general, this "sweeping away" only obtains *locally*, not *globally* if the intrinsic curvature is nonzero.]

p. 14, ll. 4-9 [Having just chosen to "sweep away" curvature at the given point, then near that point, the surface is approximated by a tangent plane, whose derivatives with respect to x and y are linear, which is satisfied by l. 4, for which $\partial z/\partial x = T^o x + U^o y$ + higher order terms. Then Gauss removes the cross term $U^o xy$ by rotating the coordinate axes in the same way that in analytic geometry a general quadratic expression can be put in standard form. If we rotate the coordinates by an angle M, then $x = x' \cos M - y' \sin M$, $y = x' \sin M + y' \cos M$. Substituting this into the expression for z in l. 4, the coefficient of $x'y'$ will be $-\frac{1}{2}(T^o - V^o)$ $2 \sin M \cos M + U^o (\cos^2 M - \sin^2 M) = -\frac{1}{2}(T^o - V^o) \sin 2M + U^o \cos 2M$, which vanishes if $\sin 2M/\cos 2M = \tan 2M = 2U^o/(T^o - V^o)$, as Gauss states.]

p. 14, §8I [Since the curvature in the x direction is $\partial^2 z/\partial x^2 = T$, then the radius of curvature in that direction is $1/T$, and likewise for the y direction (II).]

p. 15, §8V [The relation $k = TV$ follows from Gauss's equation at the end of §7 (p. 13, l. 4b) because we have arranged that $U = 0$ and we are taking the limit $t, u \to 0$ as we approach the point A. The "two extreme radii of curvature" are sometimes also called the *principal radii of curvature*. The concept of mean curvature was introduced by Sophie Germain (1831; see p. 52, l. 6b, and Kline 1972, 884).]

p. 15, l. 2b [Because $t = \partial z/\partial x$ and $dW = P\ dx + Q\ dy + R\ dz = 0$ (p. 7, l. 14), taking $\partial W/\partial x = P + R(\partial z/\partial x)gh$ and so $t = -P/R$.]

p. 16, ll. 7-9 [Recall the definitions of T, U, V on p. 13, l. 4.]

p. 16, l. 3b [This follows from the equation $X\ dx + Y\ dy + Z\ dz = 0$ (p. 7, l. 7) and the solutions for X, Y, Z given on p. 8., l. 11.]

p. 17, l. 9f [The total differential of t is $dt = (\partial t/\partial p)\ dp + (\partial t/\partial q)\ dq$ and likewise for u. Then

Gauss uses the expressions at the top of the page to calculate $\partial t/\partial p$, $\partial t/\partial q$ along with the expressions he has just found for dp, dq. The following six formulas follow from the definitions of A, B, C, (p. 16, ll. 7b-5b), a, b, c (p. 7, ll. 3b-1b), and α, β, γ, ...(p. 16, ll. 11b-9b), noting that $\partial a/\partial p = \partial^2 x/\partial p^2 = \alpha$, $\partial b/\partial p = \partial^2 y/\partial p^2 = \beta$, etc.]

p. 17, ll. 6b-1b [Since $C^3\,dt = C^3 T\,dx + C^3 U\,dy$, we can compare this form with that given for $C^3\,dt$ by the total differential of t (l. 10 above), then equate the coefficients of dx and dy separately to T and U. In the case of T, this gives an expression that explicitly shows all the terms in Gauss's equation at the bottom of p. 17 except those containing factors of B. To obtain these, we need to substitute the definitions of A and C into the remaining terms and rearrange so as to show the factors of $B = ca' - ac'$. Exactly similar procedures yield the following equations for U and V.]

p. 20, ll.5-7 [Let us compare this with the modern tensor notation used in general relativity. If we express the infinitesimal distance $ds^2 = g_{11}du^2 + 2g_{12}du dv + g_{22}dv^2 = \Sigma_{ij} g_{ij}\, dx^i\, dx^j$ where g_{ij} is the *metric tensor* whose determinant $g = g_{11}\, g_{22} - (g_{12})^2$, then Gauss's expression is the equivalent of the modern statement that $R_{1212}/g = \partial\Gamma_{221}/\partial u - \partial\Gamma_{121}/\partial v + \Gamma_{11i}\,\Gamma_{12}^{\ i} - \Gamma_{11i}\,\Gamma_{22}^{\ i}$. Here R_{ijkl} is the Riemann tensor, Γ_{ijk} the affine connection (which is in metric spaces the negative of the Christoffel symbol of the second kind), and the repeated index i is summed over (see Reich 1981, 101).]

p. 20, ll. 7b-5b [By "development" Gauss means an isometry: any (one-to-one) transformation that preserves distances (metric relations). If so, then the distance function in the new coordinates x',y',z', when reduced to Gaussian coordinates again, must agree with the original distance function. Therefore $\sqrt{(E\,dp^2 + 2F\,dp\,dq + G\,dq^2)} = \sqrt{(E'\,dp^2 + 2F'\,dp\,dq + G'\,dq^2)}$ which in general must require $E = E'$, $F = F'$, $G = G'$. Thus Gauss's amazingly elegant and simple proof crucially requires his coordinates p, q and the assumption of distance-preserving transformations (isometries).]

p. 21, l. 8 [This is just the statement that the measure of curvature k must vanish if $TV - U^2 = 0$, using the expression for k on p. 13, l. 3b and the definitions of T, U, V on l. 4 of that page.]

p. 21, l. 1b [This is the *parametric representation* of the curve in terms of the parameter w.]

p. 22, ll. 3-10 [The length of the line is given by the integral of the distance along it, as in l. 3. Gauss uses the notation for the calculus of variations as Euler had already established it (and as it remains), so that the variation of the line's length is δ applied to the integral in l. 3. Since δ applies only to variations in x, y, z, it leaves w unaffected and we can write its effect on the integrand using the chain rule:
$\delta\sqrt{((dx)^2 + (dy)^2 + (dz)^2)} = (2\,dx\,d\delta x + 2\,dy\,d\delta y + 2\,dz\,d\delta z)/2\sqrt{((dx)^2 + (dy)^2 + (dz)^2)}$, as in l. 7. Then Gauss applies integration by parts: $\int u\,dv = uv - \int v\,du$, where for the first term $u = dx$

$/\sqrt{((dx)^2 + (dy)^2 + (dz)^2)}$, $dv = d\delta x$, and likewise for the exactly similar y and z terms, thus yielding ll. 9-10.]

p. 22, ll. 12b-1b [By "well-known rules," Gauss means that, since the variations must satisfy $P\,\delta x + Q\,\delta y + R\,\delta z = 0$ in order to remain on the given surface $P\,dx + Q\,dy + R\,dz = 0$, then so too must the integrand in l. 10, which implies that the coefficients of δx, δy, δz in the integrand must also satisfy this equation and hence must be proportional to P, Q, R respectively, as given in l. 10b. Recall that λ is the point on the sphere that represents the direction of the line element in the sense that λ is the zenith point of the "horizon" given by the line element being integrated over. L is then the normal to the curved surface at the point on the surface in question; if $dr = \sqrt{((dx)^2 + (dy)^2 + (dz)^2)}$, then ξ, η, ζ are the direction cosines of λ, so that $\xi = dx/dr$, etc. P, Q, R are proportional to X, Y, Z respectively because of a result from analytic geometry: any normal vector to a surface $P\,dx + Q\,dy + R\,dz = 0$ is proportional to $Px + Qy + Rz$, hence so too is $Xx + Yy + Zz$.]

p. 24, l. 3 [The last term on the right-hand side of this equation as required by the chain rule corresponds to $(\partial x/\partial r)\,(\partial^2 x/\partial\varphi\partial r) = \frac{1}{2}(\partial x/\partial r)\,(\partial/\partial\varphi)\,(\partial x/\partial r)^2$, plus corresponding terms in y and z.]

p. 24, ll. 9bf [Gauss is considering the triangle

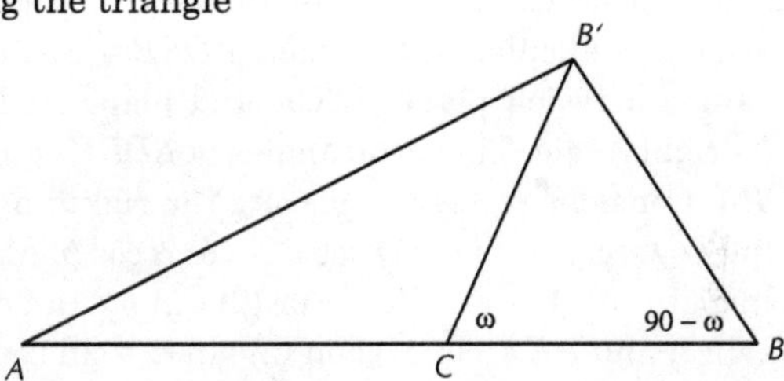

in which the angle at C is ω because C is placed such that $CB' = BC \cos\omega$. Then p. 25, l. 2 shows that, if $\omega > 0$, then line ACB' would be shorter than the assumed shortest line AB', which is absurd, so $\omega = 0$ and the lines connecting the extremities are thus perpendicular.]

p. 25, §17 [See the notes given on p. 54; from this point on, the notes by Morehead and Hiltebeitel following p. 54 offer their detailed commentary.]

p. 30, ll. 11b-7b [This is often referred to as the "Gauss-Bonnet theorem"; see Bonnet 1852.]

p. 43 [For commentary and references regarding these surveying results and their implications, see the introduction.]

Spherical triangles are composed of three arcs of great circles. We begin by considering a *right spherical triangle* ΔABC having a right angle at C and acute angles at A and B. We will denote by a the arc opposite angle A, and likewise for the other angles:

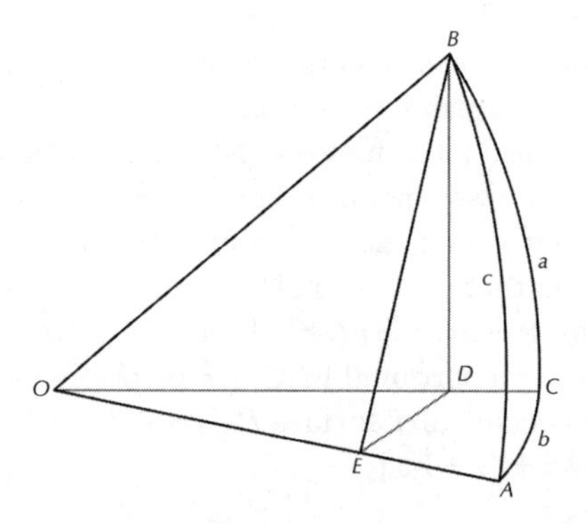

Note that, since arcs such as a are measured around a circle, they can be measured by the same degree or radian measure used for angles like A. Note also the *trihedral angle O-ACB* subtended by the whole triangle ABC as seen from the center of the sphere, which is assumed to have unit radius ($OA = OB = OC = 1$). In this figure, we have also drawn a plane through point B perpendicular to line OA, intersecting the trihedral angle in ΔBDE. Because BE and DE are perpendicular to OA, then $\angle DEB = \angle A$, each having the same measure as the dihedral angle between plane $ABOE$ and plane $ACDOE$. Likewise, $\angle DBE = \angle B$ and $\angle EDB = \angle C =$ right angle. Thus, the angles of ΔDBE equal the corresponding angles of the spherical ΔCBA. Consider the arc c opposite the right angle C. Since $\cos c = OE/OB = OE = OD \cos b$ and $OD = \cos a$, then (1) $\cos c = \cos a \cos b$. Also note that $\cos A = DE/BE = (OE \tan b)/(OE \tan c)$, so $\cos A = \tan b \cot c$ or (2) $\tan b = \tan c \cos A$. Finally, note that since $\sin a = BD = BE \sin A$ and $BE = \sin c$, then (3) $\sin a = \sin c \sin A$ and (by exactly similar arguments) (4) $\sin b = \sin c \sin B$.

Now let us consider a

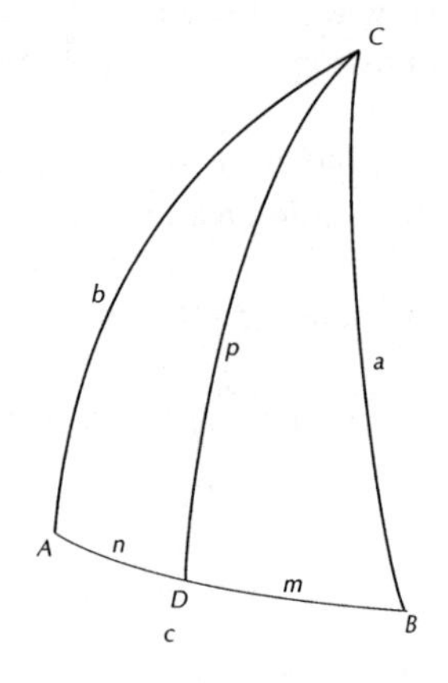

general (nonright) spherical ΔABC, here seen looking down on the surface of the sphere: Here we have also drawn an arc CD perpendicular to AB. Applying (1) to the right ΔCDB just formed by gives (5): $\cos a = \cos p \cos m = \cos p \cos (c - n) = \cos p \cos c \cos n + \cos p \sin c \sin n$, using the addition formula for the cosine. Likewise, applying (1) to triangle ACD gives (6) $\cos b = \cos p \cos n$, from which it follows that $\cos p = \cos b \sec n$ and then (multiplying both sides by $\sin n$) $\cos p \sin n = \cos b \tan n$. Using (2), $\tan n = \tan b \cos A$ and substituting all these expressions into (5) gives the *law of cosines for spherical triangles*,

$\cos a = \cos b \cos c + \sin b \sin c \cos A,$
$\cos b = \cos c \cos a + \sin c \sin a \cos B,$
$\cos c = \cos a \cos b + \sin a \sin b \cos C,$

the latter two results from exactly analogous arguments as above (or simply applying to our result cyclic permutation $a \to b \to c \to a$, $A \to B \to C \to A$).

Also note that, in our same spherical ΔABC, (3) implies that $\sin p = \sin a \sin B$ and (4) implies that $\sin p = \sin b \sin A$. Dividing these (and applying similar arguments also to arc c) gives

$\sin a / \sin A = \sin b / \sin B = \sin c / \sin C,$

the *law of sines for spherical triangles.*

In the above, we have assumed that $\angle A$ and $\angle B$ are each less than one right angle; indeed, a right spherical triangle can have three right angles. If we redo the figures to include the case that angles A or B is obtuse, we get the same laws of cosines and sines as above. For further review of basic spherical trigonometry, see Klaf 2005.

Adler, Ronald. Maurice Bazin, and Menahem Schiffer. *Introduction to General Relativity*, second edition. New York: McGraw-Hill, 1975.

Bliss, Gilbert Ames. "Generalizations of Geodesic Curvature and a Theorem of Gauss Concerning Geodesic Triangle," *American Journal of Mathematics* **37**, 1-15 (1915).

Bonnet, Ossian. "Sur la Théorie mathématique des Cartes géographiques." *Journal de mathématiques pures et appliquées* **17**, 301-340 (1852).

Bonola, Roberto. *Non-Euclidean Geometry: A Critical and Historical Study of its Development*. Dover, 1955.

Bottazzini, Umberto. "Geometry and 'metaphysics of space' in Gauss and Riemann," in Poggi and Bossi 1994, 15-29.

Boyer, Carl B. *A History of Mathematics*, second edition, revised by Uta C. Merzbach. New York: Wiley, 1991.

Breitenberger, Ernst. "Gauss's Geodesy and the Axiom of Parallels," *Archive for the History of Exact Science* **31**, 273-289 (1984).

Bühler, W. K. *Gauss: A Biographical Study*. Berlin: Springer-Verlag, 1981.

Chikara, Sasaki *et al.* (editors). *Intersection of History and Mathematics*. Boston: Birkhäuser, 1994.

Cooke, Roger. *The History of Mathematics: A Brief Course,* second edition. New York: Wiley-Interscience, 2005.

Coolidge, Julian Lowell. *A History of Geometrical Methods*. Dover, 1963, 355-376.

Daniels, Norman. "Lobachvsky: Some Anticipations of Later Views on the Relation between Geometry and Physics," *Isis* **66**, 75-87 (1975).

Demidov, Sergei S., M. Folkerts, D. E. Rowe, C. J. Scriba (editors). *Amphora: Festschrift für Hans Wussing*. Basel : Birkhäuser, 1992.

Dombrowski, Peter. *150 Years After Gauss'* Disquisitiones generales circa superfices curva. Paris: Société Mathématique de France, 1979.

Duistermat, J. J. "Gauss on Surfaces," in Monna 1978, 32-41.

Dunnington, G. Waldo. "Note on Gauss' Triangulations," *Scripta Mathematica* **20**, 108-109 (1953).

Dunnington, G. Waldo. *Carl Friedrich Gauss: Titan of Science,* with additional material by Jeremy Gray and Fritz-Egbert Dohse. Washington, D.C.: Mathematical Association of America, 2004.

Euler, Leonhard. *Opera omnia*. Basel: Birkhäuser, 1975.

Ewald, William. *From Kant to Hilbert: A Source Book in the Foundations of Mathematics.* Oxford: Clarendon Press, 1996.

Forbes, Eric G. "The astronomical work of Carl Friedrich Gauss," *Historia Mathematica* 5, 167-181 (1978).

Franks, Grant. "The Pirouette in Gauss's *General Investigation of Curved Surfaces.*" Unpublished essay, 2002.

Gauss, Karl Friedrich. *General Investigations of Curved Surfaces of 1827 and 1825*, translated with notes and bibliography by James Caddall Morehead and Adam Miller Hiltebeitel. Princeton: Princeton University Library, 1902.

Gauss, Carl Friedrich. *Inaugural Lecture on Astronomy and Papers on the Foundations of Mathematics*, translated and edited by G. Waldo Dunnington. Baton Rouge: Lousiana State University Press, 1937.

Gauss, Carl Friedrich. *Werke*. Hildesheim: Georg Olms, 1973.

Gauss, Carl Friedrich Gauss., *General Investigations of Curved Surfaces*, with introduction by Richard Courant. Hewlett, NY: Raven Press, 1965.

Gauss, Karl Friedrich Gauss, *Theory of the Motion of the Heavenly Bodies Moving about the Sun in Conic Sections: A Translation of* Theoria Motus, translated by Charles Henry Davis. Dover, 2004.

Germain, Sophie. "Mémoire sur la courbure des surfaces," *Journal für Mathematik* **7**, 1-29 (1831).

Goe, George and B. L. van der Waerden, "Comments on Miller's 'The Myth of Gauss' Experiment on the Euclidean Nature of Physical Space'," with a reply by Arthur I. Miller, in *Isis* **65**, 83-87 (1974).

Golos, Ellery B. *Foundations of Euclidean and Non-Euclidean Geometry.* New York: Holt, Rinehart, and Winston, 1968.

Grattan-Guinness, Ivor (editor). *Landmark Writings in Western Mathematics, 1640-1940.* Amsterdam: Elsevier, 2005.

Gray, Jeremy. "Non-Euclidean Geometry – A Re-Interpretation," *Historia Mathematica* **6**, 236-258 (1979)

Gray, Jeremy. *Ideas of Space: Euclidean, Non-Euclidean, and Relativistic*, second ed. (Oxford: Clarendon Press, 1989),

Gray, Jeremy. "Complex Curves: Origins and Intrinsic Geometry," in Chikara 1994, 39-50.

Gray, Jeremy (editor). *The Symbolic Universe: Geometry and Physics 1890-1930.* Oxford: Oxford University Press, 1999.

Gray, Jeremy. *János Bolyai, Non-Euclidean Geometry, and the Nature of Space.* Cambridge, Mass.: Burndy Library Publications, 2004.

Greenberg, Marvin Jay. *Euclidean and Non-Euclidean Geometries: Development and History*, second edition. New York: W. H. Freeman, 1974.

Grossmann, W. "Gauß' geodätische Tätigkeit im Rahmen zeitgenössischer Arbeiten," *Zeitschrift für Vermessungswesen* **80**, 371-384 (1955),

Hall, Tord. *Carl Friedrich Gauss*, translated by Albert Froderberg. Cambridge, Mass.: MIT Press, 1970.

Hilbert, David. *Foundations of Geometry*, translated by Leo Unger, tenth edition. La Salle, Ill.: Open Court, 1959.

Hoppe, Edmund. "C. F. Gauss und der euklidische Raum." *Die Naturwissenschaften* **13**, 743-744 (1925).

Jammer, Max. *Concepts of Space: The History and Theories of Space in Physics*, second edition. Cambridge, Mass: Harvard University Press, 1969.

Klaf, A. Albert, *Trigonometry Refresher*. Dover, 2005.

Klein, Felix. *Development of Mathematics in the Nineteenth Century*, translated by M. Ackerman. Brookline, Mass: Math Sci Press, 1979.

Kline, Morris. *Mathematical Thought from Ancient to Modern Times.* New York: Oxford University Press, 1972).

Lambert, J. H. *Anmerkungen und Zusätze zur Entwerfung der Land- und Himmelscharten.* Leipzig: Engelmann, 1896.

Lanczos, Cornelius. *Albert Einstein and the Cosmic World Order.* New York: Wiley, 1965.

Legendre, Adrien Marie. *Éléments de géometrie, avec des notes.* Paris: Firmin Didot, 1794.

Lützen, Jesper. "Interactions between Mechanics and Differential Geometry in the 19th Century," *Archive for the History of Exact Science* **49**, 1-72 (1995).

Maxwell, James Clerk. *A Treatise on Electricity and Magnetism*, third edition. Dover, 1954.

Merzbach, Uta C. *Carl Friedrich Gauss: A Bibliography.* Wilmington, Del.: Scholarly Resources, 1984.

Meschkowski, H. *Noneuclidean Geometry*, translated by Abe Schnitzer. New York: Academic Press, 2000.

Miller, Arthur I. "The Myth of Gauss' Experiment on the Euclidean Nature of Physical Space," *Isis* **63**, 345-348 (1972.

Monastyrsky, Michael. *Riemann, Topology, and Physics*, translated by James King and Victoria King. Boston: Birkhäuser, 1987.

Monna, A. F. (editor). *Carl Friedrich Gauss 1777-1855: Four Lectures on His Life and Work.* Utrecht: Communications of the Mathematical Institute, 1978.

Pesic, Peter and Stephen P. Boughn. "The Weyl-Cartan theorem and the naturalness of general relativity," *European Journal of Physics* **24**, 261-266 (2003).

Poggi, Stefano and Maurizio Bossi (editors). *Romanticism in Science: Science in Europe, 1790-1840.* Dordrecht : Kluwer Academic, 1994.

Reich, Karin. "Die Geschichte der Differentialgeometrie von Gauss bis Riemann (1828-1868)." *Archive for History of Exact Science* **11**, 273-382 (1973).

Reich, Karin. *Carl Friedrich Gauss 1777/1977.* Munich: Moos, 1977.

Reich, Karin. "Geodäsie und Differentialgeometrie," in Schneider 1981, 85-111.

Reichardt, Hans. *Gauss und die Anfänge der nicht-euklidischen Geometrie.* Leipzig: Teubner, 1985.

Rindler, Wolfgang. "General relativity before special relativity: An unconventional overview of relativity theory," *American Journal of Physics* **62**, 887-893 (1994).

Sheynin, Oscar. "C. F. Gauss and Geodetic Observations," *Archive for History of Exact Sciences* **46**, 253-283 (1994).

Schmidt, Klaus-Jürgen. "Ergänzungen zur Harztriangulation 1833, nach Gauss' und Hartmanns Briefwechsel mit Oberbergrat Albert Klaus-Jürgen," *Mitteilungen, Gauss-Gesellschaft* **28**: 55-73 (1991).

Schneider, Ivo (editor). *Carl Friedrich Gauß (1777-1855): Sammelband von Beiträgen zum 200. Geburtstag von C. F. Gauß.* Munich: Minerva, 1981.

Scholz, Erhard. "Gauss und die Begründung der 'höheren' Geodäsie," in Demidov 1992, 631-647.

Smith, David Eugene (editor). *A Source Book in Mathematics.* Dover, 1959.

Spivak, Michael. *A Comprehensive Introduction to Differential Geometry*, second edition. Berkeley: Publish or Perish, 1979.

Wall, C. T. C. *A Geometric Introduction to Topology.* Dover, 1993.

Waterhouse, William C. "Gauss on Infinity," *Historia Mathematics* **6,** 430-436 (1979).

Weyl, Hermann. *Philosophy of Mathematics and Natural Science*, translated by Olaf Helmer. Princeton: Princeton University Press, 1949, 130-138.

Wilson, Curtis. "Carl Friedrich Gauss, Book on Celestial Mechanics (1809)," in Grattan-Guinness 2005, 316-328.

Worbs, E. *Carl Friedrich Gauss Ein Lebensbild.* Leipzig: Koehler & Amelang, 1955.

Wussing, Hans. *Carl Friedrich Gauss*, second edition. Leiptzig: Teubner, 1976.

Yaglom, I. M. *Felix Klein and Sophus Lie: Evolution of the Idea of Symmetry in the Nineteenth Century.* Boston: Birkhäuser, 1988, 46-70.

Zormbala, Konstantina. "Gauss and the Definition of the Plane Concept in Euclidean Elementary Geometry." *Historia Mathematica* **23**, 418-436 (1996).